Lecture Notes in Computer Science   10150

Commenced Publication in 1973
Founding and Former Series Editors:
Gerhard Goos, Juris Hartmanis, and Jan van Leeuwen

More information about this series at http://www.springer.com/series/7407

Inês Dutra · Rui Camacho
Jorge Barbosa · Osni Marques (Eds.)

High Performance Computing for Computational Science – VECPAR 2016

12th International Conference
Porto, Portugal, June 28–30, 2016
Revised Selected Papers

 Springer

Editors
Inês Dutra
University of Porto
Porto
Portugal

Rui Camacho
University of Porto
Porto
Portugal

Jorge Barbosa
University of Porto
Porto
Portugal

Osni Marques
Lawrence Berkeley National Laboratory
Berkeley, CA
USA

ISSN 0302-9743   ISSN 1611-3349 (electronic)
Lecture Notes in Computer Science
ISBN 978-3-319-61981-1   ISBN 978-3-319-61982-8 (eBook)
DOI 10.1007/978-3-319-61982-8

Library of Congress Control Number: 2017946688

LNCS Sublibrary: SL1 – Theoretical Computer Science and General Issues

Printed on acid-free paper

This Springer imprint is published by Springer Nature
The registered company is Springer International Publishing AG
The registered company address is: Gewerbestrasse 11, 6330 Cham, Switzerland

Preface

The International Meeting on High-Performance Computing for Computational Science (VECPAR) is a biannual conference and is the premier venue for presenting and discussing the latest research and practice in high-end computer modeling and complex systems. The audience and participants of VECPAR are researchers and students in academic departments, government laboratories, and industrial organizations. There is a permanent website for the conference series at http://vecpar.fe.up.pt. In this 2016 edition, the conference went back to Porto, where it originated. Previous editions of VECPAR were held in Oregon (USA, 2014), Kobe (Japan, 2012), Berkeley (USA, 2010), Toulouse (France, 2008), Rio de Janeiro (Brazil, 2006), Valencia (Spain, 2004), and Porto (Portugal, 2002, 2000, 1998, 1996 and 1993).

This VECPAR edition had a very exciting program with 20 papers accepted from 10 different countries. The acceptance rate was 51%, repeating the pattern of previous editions. We had a varied selection of paper subjects ranging from distributed to shared and hybrid parallel algorithms and systems, but with a strong focus on computational science applications. Examples are epidemic modeling and word searching. Parallel libraries was also a popular subject this year. Studies of parallel platforms range from clouds to multi-core, many-core, and GPUs. The conference also had the contribution of invited talks given by four prominent speakers (Prof. Omar Ghattas, "Scalable Algorithms for Bayesian Inference of Large-Scale Models from Large-Scale Data", Prof. Bruno Schulze, "HPC as a Service", Prof. Mateo Valero, "Runtime Aware Architectures", and Prof. Luc Giraud, "Numerical Resiliency in Iterative Linear Algebra Calculation"), two workshops: the Workshop on "Big Data and Deep Learning in HPC," on June 30; and the Workshop on "Computational Challenges for Climate Modelling and Weather Prediction," on July 1; as well as a crash course on Multithreading and Vectorization on Intel®Xeon™ and Intel®Xeon Phi™ Architectures using OpenMP. In the social program, we had a welcome reception and a fantastic social dinner including Port wine tasting.

A co-organized summer school on "Advanced Scientific Computing" was organized in Braga, during the previous week, in collaboration with the Texas Advanced Computing Center (TACC), at Austin, Texas, USA.

The most significant contributions to VECPAR 2016 are made available in the present book, edited after the conference and after a second review of all accepted papers that were presented.

The paper submission and selection processes were managed via the EasyChair conference management system. The website of the conference is maintained by the Faculty of Engineering of the University of Porto.

The success of the VECPAR conference and its long life are a result of the collaboration of many people. For the 2016 edition, we would like to thank Alexandra Ferreira and Isabel Gonçalves, our very efficient secretaries in the Department of Computer Science of the University of Porto. We would also like to thank all the organizers, reviewers, and authors for their fantastic work and for meeting tight deadlines.

December 2016

Jorge Barbosa
Rui Camacho
Inês Dutra
Osni Marques

Organization

VECPAR 2016, the 12th edition of the VECPAR series of conferences, was organized by the Department of Computer Science, Faculty of Sciences of the University of Porto, Porto, Portugal.

Executive Committee

Conference Chairs

Jorge Barbosa	University of Porto, Portugal
Inês Dutra	University of Porto, Portugal
Rui Camacho	University of Porto, Portugal
Osni Marques	LBL, USA

Workshops Chair

João Manuel R.S. Tavares	University of Porto, Portugal

Publicity Chair

Carlos Ferreira	ISEP, Portugal

Web Chair

Vítor Carvalho	University of Porto, Portugal

Steering Committee

Osni Marques (Chair)	Lawrence Berkeley National Laboratory, USA
Alvaro Coutinho	COPPE/UFRJ, Brazil
Michel Daydé	ENSEEIHT, France
Jack Dongarra	University of Tennessee, USA
Inês Dutra	University of Porto, Portugal
Kengo Nakajima	University of Tokyo, Japan
Sameer Shende	University of Oregon, USA

Scientific Committee

Cláudio Amorim, Brazil
Filipe Araújo, Portugal
Cristiana Bentes, Brazil
Cristina Boeres, Brazil
Xing Cai, Norway

João Cardoso, Portugal
Lucia Catabriga, Brazil
Alvaro Coutinho, Brazil
Yifeng Cui, USA
Michel Daydé, France

Pedro Diniz, USA
Jorge González-Domínguez, Spain
Tingxing Dong, USA
Lúcia Drummond, Brazil
Felipe M.G. França, Brazil
Akihiro Fujii, Japan
Claudio Geyer, Brazil
Laura Grigori, France
Ronan Guivarch, France
Abdelkader Hameurlain, France
Antonio J. Tomeu-Hardasmal, Spain
Toshiyuki Imamura, Japan
Alexandru Iosup, The Netherlands
Florin Isaila, Spain
Takeshi Iwashita, Japan
Helen Karatza, Greece
Takahiro Katagiri, Japan
Jakub Kurzak, USA
Daniel Kressner, Switzerland
Stéphane Lanteri, France
Alexey Lastovetsky, Ireland
Paul Lin, USA
Jean-Yves L'Excellent, France
João Lourenço, Portugal
Piotr Luszczek, USA
Tomas Margalef, Spain
Pedro Medeiros, Portugal
Kengo Nakajima, Japan
Kenji Ono, Japan

Satoshi Ohshima, Japan
Hervé Paulino, Portugal
Maria S. Perez, Spain
Alberto Proença, Portugal
Rui Ralha, Portugal
Doallo Ramón, Spain
Vinod Rebello, Brazil
Lígia Ribeiro, Portugal
Pedro Ribeiro, Portugal
Francisco F. Rivera, Spain
Ricardo Rocha, Portugal
Paolo Romano, Portugal
Rizos Sakellariou, UK
Tetsuya Sakurai, Japan
Sameer Shende, USA
Fernando Silva, Portugal
David E. Singh, Spain
João Sobral, Portugal
A. Augusto Sousa, Portugal
Leonel Sousa, Portugal
Reiji Suda, Japan
Frederic Suter, France
Domenico Talia, Italy
Keita Teranishi, USA
Mirek Tuma, Czech Republic
Paulo Vasconcelos, Portugal
Xavier Vasseur, France
Luís Veiga, Portugal
Weichung Wang, Taiwan

Additional Reviewers

Diego Dutra, Brazil
Yulu Jia, USA
Tiago Vale, Portugal
Paulo Martins, Portugal

João Gante, Portugal
Nuno Oliveira, Portugal
Roberto R. Expósito, Spain

Sponsoring Institutions

Springer International Publishing
Porto Convention Bureau, Porto, Portugal
ParaTools, Inc., 2836 Kincaid St., Eugene, OR 97405, USA
HP Portugal, Information Technology and Services
Reitoria da Universidade do Porto, Porto, Portugal

Contents

Performance Modeling and Analysis

Low Level Support

Environments/Libraries to Support Parallelization

Invited Talks

Scalable Algorithms for Bayesian Inference of Large-Scale Models from Large-Scale Data

Omar Ghattas[1]([⊠]), Tobin Isaac[2], Noémi Petra[3], and Georg Stadler[4]

[1] Institute for Computational Engineering and Sciences,
Departments of Geological Sciences and Mechanical Engineering,
The University of Texas at Austin, Austin, USA
omar@ices.utexas.edu
[2] Computation Institute, University of Chicago, Chicago, USA
[3] School of Natural Sciences, University of California, Merced, Merced, USA
[4] Courant Institute for Mathematical Sciences, New York University, New York, USA

One of the greatest challenges in computational science and engineering today is how to combine complex data with complex models to create better predictions. This challenge cuts across every application area within CS&E, from geosciences, materials, chemical systems, biological systems, and astrophysics to engineered systems in aerospace, transportation, structures, electronics, biomedicine, and beyond. Many of these systems are characterized by complex nonlinear behavior coupling multiple physical processes over a wide range of length and time scales. Mathematical and computational models of these systems often contain numerous uncertain parameters, making high-reliability predictive modeling a challenge. Rapidly expanding volumes of observational data—along with tremendous increases in HPC capability—present opportunities to reduce these uncertainties via solution of large-scale inverse problems.

In an inverse problem, we infer unknown model parameters (e.g., coefficients, material properties, source terms, initial or boundary conditions, geometry, model structure) from observations of model outputs. The need to quantify the uncertainty in the solution of such inverse problems has attracted widespread attention in recent years. This can be carried out in a systematic manner by casting the inverse problem within the framework of Bayesian inference. In this framework, uncertain observations and uncertain models are combined with available prior knowledge to yield a probability density in the model parameters as the solution of the inverse problem, thereby providing a rational and systematic means of quantifying uncertainties in the inference of these parameters. The resulting uncertainties in model parameters are then propagated forward through models to yield predictions with associated uncertainty. Finally, given this capability to quantify uncertainties in inverse problems, one can determine the design of the observational system (e.g., location of sensors, nature of

This work was supported by AFOSR grants FA9550-12-1-0484 and FA9550-09-1-0608, DARPA/ARO contract W911NF-15-2-0121, DOE grants DE-SC0010518, DE-SC0009286, DE-11018096, DE-SC0006656, DE-SC0002710, and DE-FG02-08ER25860, and NSF grants ACI-1550593, CBET-1508713, CBET-1507009, CMMI-1028889, and ARC-0941678. Computations were performed on supercomputers at TACC, ORNL, and LLNL. We gratefully acknowledge this support.

© Springer International Publishing AG 2017
I. Dutra et al. (Eds.): VECPAR 2016, LNCS 10150, pp. 3–6, 2017.
DOI: 10.1007/978-3-319-61982-8_1

measured quantities) that maximizes the information gain from the observations (or minimizes the uncertainty in the inferred model or subsequent prediction). This is the optimal experimental design (OED) problem, which wraps an optimization problem around the Bayesian inverse problem.

The Markov chain Monte Carlo (MCMC) method has emerged as the method of choice for solving Bayesian inverse problems. Unfortunately, when the forward model is large and complex (e.g., when the model takes the form of an expensive-to-solve system of partial differential equations), and when the parameters are high-dimensional (as results from discretization of an infinite dimensional field such as an initial condition or heterogeneous material property), solution of Bayesian inverse problems via conventional MCMC is intractable. Moreover, addressing the meta-question of how to optimally obtain experimental data for such problems via solution of an OED problem is completely out of the question.

However, a number of advances over the past decade have brought the goal of Bayesian inference of large-scale complex models from large-scale complex data much closer. First, improvements in scalable forward solvers for many classes of large-scale models have made feasible numerous evaluations of model outputs for differing inputs. Second, sustained growth in HPC capabilities has multiplied the effects of the advances in solvers. Third, the emergence of MCMC methods that exploit problem structure (e.g., curvature of the posterior probability) has radically improved the prospects of sampling posterior distributions for inverse problems governed by expensive models. And fourth, recent exponential expansions of observational capabilities have produced massive volumes of data from which inference of large computational models can be carried out.

To overcome the prohibitive nature of Bayesian methods for high-dimensional inverse problems governed by expensive-to-solve PDEs, we exploit the fact that, despite the large size of observational data, they typically provide only sparse information on model parameters. This implicit dimension reduction is provided by low rank approximations of the Hessian of the data misfit functional, which is typically a compact operator due to ill-posedness of the inverse problem. A low rank approximation of the Hessian can be extracted efficiently in a matrix-free manner (without forming the Hessian) by a Lanczos [8, 14] or randomized SVD [4, 5, 12, 15, 21] method, requiring a number of matrix-vector products that scales only with the rank of the Hessian, and not the parameter dimension. Moreover, the rank reflects how informative the data are, i.e., how many directions in parameter space are informed by the data. Finally, each Hessian-vector product can be computed using just a pair of linearized forward/adjoint PDE solves [4, 5, 8, 9, 12, 14–17, 21, 22].

We have applied the methodology described above (for exploiting the geometric structure of the posterior) to geophysical inverse problems arising in ice sheet flow, seismic wave propagation, mantle convection, atmospheric transport, poromechanics, and subsurface flow. We are able to substantially reduce the effective parameter dimension (often by three orders of magnitude) at a cost, measured in (linearized) forward/adjoint PDE solves, that is independent of both the parameter and data dimensions [4, 5, 8, 9, 12, 14, 15, 20, 21].

For linearized Bayesian analysis of nonlinear inverse problems, the Hessian evaluated at the point in parameter space that maximizes the posterior (i.e., the MAP point) completely characterizes the uncertainty in inferred parameters. One can build on this idea to solve optimal experimental design problems at a cost that also does not scale with the parameter or data dimensions [1–3]. For nonlinear Bayesian inverse problems, the Hessian varies from point to point. However the low rank Hessian approximation machinery described above can still be exploited to accelerate MCMC sampling, by serving as an inverse covariance approximation for a Gaussian proposal that is tailored to the local curvature of the posterior [14,15] (this is known as the stochastic Newton method).

The most complex inverse problem for which we have carried out Bayesian inversion involves ice sheet flow [12,15,16,22]. The flow of ice from polar ice sheets such as Antarctica and Greenland is the primary contributor to projected sea level rise in the 21st century. The ice is modeled as a creeping, viscous, incompressible, non-Newtonian, shear-thinning fluid, for which we have developed custom scalable parallel solvers [13,18,19] on adaptively refined forest-of-octree meshes [6,7,10,11], the combination of which has scaled to hundreds of billions of unknowns on up to 1.6 million cores [4,6,18]. One of the main difficulties faced in modeling ice sheet flow is the unknown spatially-varying Robin boundary condition that describes the resistance to sliding at the base of the ice. Satellite observations of the surface ice flow velocity can be used to infer this uncertain basal boundary condition. We have solved this ill-posed inverse problem using the (linearized) Bayesian inference machinery described above, which allows us to infer not only the unknown basal sliding parameters, but also the associated uncertainty [12]. We have demonstrated that the number of required forward solves is independent of the parameter dimension, data dimension, and number of processor cores. The largest Bayesian inverse problem solved has over one million uncertain parameters.

References

1. Alexanderian, A., Petra, N., Stadler, G., Ghattas, O.: A-optimal design of experiments for infinite-dimensional Bayesian linear inverse problems with regularized ℓ_0-sparsification. SIAM J. Sci. Comput. **36**(5), A2122–A2148 (2014)
2. Alexanderian, A., Petra, N., Stadler, G., Ghattas, O.: A fast and scalable method for A-optimal design of experiments for infinite-dimensional Bayesian nonlinear inverse problems. SIAM J. Sci. Comput. **38**(1), A243–A272 (2016)
3. Alexanderian, A., Gloor, P., Ghattas, O.: On Bayesian A- and D-optimal experimental designs in infinite dimensions. Bayesian Anal. **11**(3), 671–695 (2016)
4. Bui-Thanh, T., Burstedde, C., Ghattas, O., Martin, J., Stadler, G., Wilcox, L.C.: Extreme-scale UQ for Bayesian inverse problems governed by PDEs. In: Proceedings of IEEE/ACM SC12 (2012)
5. Bui-Thanh, T., Ghattas, O., Martin, J., Stadler, G.: A computational framework for infinite-dimensional Bayesian inverse problems. Part I: The linearized case, with applications to global seismic inversion. SIAM J. Sci. Comput. **35**(6), A2494–A2523 (2013)
6. Burstedde, C., Ghattas, O., Gurnis, M., Isaac, T., Stadler, G., Warburton, T., Wilcox, L.C.: Extreme-scale AMR. In: Proceedings of ACM/IEEE SC 2010 (2010)

7. Burstedde, C., Wilcox, L.C., Ghattas, O.: p4est: Scalable algorithms for parallel adaptive mesh refinement on forests of octrees. SIAM J. Sci. Comput. **33**(3), 1103–1133 (2011)

8. Flath, H.P., Wilcox, L.C., Akcelik, V., Hill, J., van Bloemen, B., Ghattas, O.: Fast algorithms for Bayesian uncertainty quantification in large-scale linear inverse problems based on low-rank partial Hessian approximations. SIAM J. Sci. Comput. **33**(1), 407–432 (2011)

9. Hesse, M., Stadler, G.: Joint inversion in coupled quasistatic poroelasticity. J. Geophys. Res. Solid Earth **119**, 1425–1445 (2014)

10. Isaac, T., Burstedde, C., Ghattas, O.: Low-cost parallel algorithms for 2:1 octree balance. In: International Parallel and Distributed Processing Symposium (IPDPS 2012), pp. 426–437. IEEE Computer Society (2012)

11. Isaac, T., Burstedde, C., Wilcox, L.C., Ghattas, O.: Recursive algorithms for distributed forests of octrees. SIAM J. Sci. Comput. **37**(5), C497–C531 (2015)

12. Isaac, T., Petra, N., Stadler, G., Ghattas, O.: Scalable and efficient algorithms for the propagation of uncertainty from data through inference to prediction for large-scale problems, with application to flow of the Antarctic ice sheet. J. Comput. Phys. **296**(1), 348–368 (2015)

13. Isaac, T., Stadler, G., Ghattas, O.: Solution of nonlinear Stokes equations discretized by high-order finite elements on nonconforming and anisotropic meshes, with application to ice sheet dynamics. SIAM J. Sci. Comput. **37**(6), B804–B833 (2015)

14. Martin, J., Wilcox, L.C., Burstedde, C., Ghattas, O.: A Stochastic Newton MCMC method for large-scale statistical inverse problems with application to seismic inversion. SIAM J. Sci. Comput. **34**(3), A1460–A1487 (2012)

15. Petra, N., Martin, J., Stadler, G., Ghattas, O.: A computational framework for infinite-dimensional Bayesian inverse problems: Part II: Stochastic Newton MCMC with application to ice sheet flow inverse problems. SIAM J. Sci. Comput. **36**(4), A1525–A1555 (2014)

16. Petra, N., Zhu, H., Stadler, G., Hughes, T.J.R., Ghattas, O.: An inexact Gauss-Newton method for inversion of basal sliding and rheology parameters in a nonlinear Stokes ice sheet model. J. Glaciol. **58**(211), 889–903 (2012)

17. Ratnaswamy, V., Stadler, G., Gurnis, M.: Adjoint-based estimation of plate coupling in a non-linear mantle flow model: theory and examples. Geophys. J. Int. **202**(2), 768–786 (2015)

18. Rudi, J., Malossi, A.C.I., Isaac, T., Stadler, G., Gurnis, M., Staar, P.W.J., Ineichen, Y., Bekas, C., Curioni, A., Ghattas, O.: An extreme-scale implicit solver for complex PDEs: highly heterogeneous flow in earth's mantle. In: Proceedings of IEEE/ACM SC 2015 (2015)

19. Rudi, J., Stadler, G., Ghattas, O.: Weighted BFBT Preconditioner for Stokes Flow Problems with Highly Heterogeneous Viscosity (submitted) (2016)

20. Worthen, J., Stadler, G., Petra, N., Gurnis, M., Ghattas, O.: Towards adjoint-based inversion for rheological parameters in nonlinear viscous mantle flow. Phys. Earth Planet. Inter. **234**, 23–34 (2014)

21. Zhu, H., Li, S., Fomel, S., Stadler, G., Ghattas, O.: A Bayesian approach to estimate uncertainty for full waveform inversion using a priori information from depth migration. Geophysics **81**(5), R307–R323 (2016)

22. Zhu, H., Petra, N., Stadler, G., Isaac, T., Hughes, T.J.R., Ghattas, O.: Inversion of geothermal heat flux in a thermomechanically coupled nonlinear Stokes ice sheet model. Cryosphere **10**, 1477–1494 (2016)

Analysis of High Performance Applications Using Workload Requirements

Mariza Ferro[✉], Giacomo Mc Evoy, and Bruno Schulze

National Laboratory of Scientific Computing, Petrópolis, Brazil
{mariza,giacomo,schulze}@lncc.br

Abstract. This short paper proposes two novel methodologies for analyzing scientific applications in distributed environments, using workload requirements. The first explores the impact of features such as problem size and programming language, over different computational architectures. The second explores the impact of mapping virtual cluster resources on the performance of parallel applications.

Keywords: HPC · Scientific computing · Performance prediction · Virtualization

1 Introduction

High Performance Distributed Computing is essential to improve scientific progress in many areas of science and to efficiently deploy a number of complex scientific applications. Also, the efficient deployment of High Performance Computing applications on Clouds offers many challenges, in particular, for communication- intensive applications. Benchmarks are good for comparisons between computational architectures, but they are not the best approach for evaluating if an architecture is adequate for a set of scientific applications. In this paper, we discuss two methodologies for evaluating the impact of the underlying infrastructure on observed performance, both from physical and virtual perspectives. The first methodology begins on scientific application characteristics, and then considers how these characteristics interact with the problem size, with the programming language and finally with a specific computational architecture. The second methodology focuses on the case of distributed applications in virtual clusters by analyzing the impact of different VM profiles and placements.

2 Methodology Based on Requirements

In this methodology, the performance evaluation is made considering the characteristics of the applications that will be used in the HPC infrastructure, under conditions as real as possible. It was developed based on Operational Analysis (OA) concepts [5] from where we extract the systematic model to evaluate complex systems and to provide a decision-making process to rationally choose an

© Springer International Publishing AG 2017
I. Dutra et al. (Eds.): VECPAR 2016, LNCS 10150, pp. 7–10, 2017.
DOI: 10.1007/978-3-319-61982-8_2

architecture. Also, it was made a study about the requirements of the scientific applications, based on applications classes named Dwarfs [1]. These classes represent the behavior in terms of computational requirements. These requirements were studied, modeled and a set of parameters were defined for the methodology (Essential Elements of Analysis - EEA).

The methodology comprises a set of phases and respective steps, briefly described next. All phases and steps of the methodology are detailed in [2].

2.1 Description of Methodology Phases

The first phase is the *Definition Problem* in which the real problem and the objective of the methodology application are clearly defined. In sequence, the phase *Problem Detailing Analysis* details the user problem, searching the complete definition of requirements. It is very important here the knowledge acquired about each application, focus of the evaluation: the real problem sizes/workload executed, programming languages, applications executed sequentially or in parallel, etc. Further, the relative importance of each one is defined in a subjective way by researchers and converted in a set of numerical weights by means of Analytic Hierarchy Process (AHP). Beyond those critical issues, the Measures of Effectiveness (MOEs) and EEA are defined. A MOE of a system is a parameter that evaluates the capability of the system to accomplish its assigned goal under a given set of conditions. The *implementation* phase is where the test planning is completed, based on both aforementioned phases. The methodology endorses that the real application and workloads must be used for performance evaluation, enabling an evaluation as real as possible. However, we know that it is not always possible, for example by confidentiality or software licenses. So, in this case the real applications are mapped to a Dwarf class. The model for mapping applications to Dwarf comprises a set of rules that enable us to define the class of an application based on the EEA measured under the execution tests. Based on the classification of each application, one or more benchmarks are defined to be executed as evaluation test. The last phase is *Communication of Results*, in which data collected on tests are confronted with MOEs and the data from different providers are compared. For this phase it was developed a Gain Function (GF) that enables the decision based on quantitative and qualitative parameters about the problem of the researcher. Using MOEs and the GF, it is possible to define the operational effectiveness and suitability of the infrastructure. The GF is briefly described in Eq. 1 [3].

$$G(k) = w_d \sum_{j=1}^{n} w_j D(j,k) + w_c C_{E_k}, k = 1, \ldots, m \qquad (1)$$

For each application j, $j = 1, ..., n$, on each evaluated infrastructure E_k, $k = 1, ..., m$, the execution time $t(j,k)$ is measured. For each application j it is assigned a weight w_j. Also, for each architecture is considered its cost c_k. Let w_c and w_d be the weights for cost and performance. From those operational values, the GF enables to consider the performance (execution time) of each scientific application for each architecture evaluated.

3   Multi-dimensional Analysis on Virtual Clusters

This methodology proposes the utilization of Canonical Correlation Analysis (CCA) to find optimal virtual cluster settings of an application, accounting for its communication pattern. It is built upon three sources of information:

1. Characteristics about how the virtual cluster is defined and deployed;
2. Characteristics of the performance of the target application;
3. Characteristics about the nature of the workload using Dwarfs.

Extracting Characteristics: The Cluster Placement [4] was proposed to address the limitations of current descriptions of virtual clusters. Most representations focus solely on the dimensions of the virtual cluster. These elements can be directly observed by a parallel application running on the cluster. With our proposed model, it is not only possible to determine which VMs execute on which physical machine, but also know how each virtual core is mapped to underlying hardware using virtual core pinning (or lack thereof). This enriched information allow us to map virtualization characteristics to performance more effectively.

In order to understand the effect that Cluster Placement exerts on the performance of an application, we developed the VESPA (Virtualized Experiments for Scientific Parallel Applications) framework that manages the systematic execution of the application along several scenarios with different Cluster Placements. Executions were performed in a controlled environment to isolate resulting variability to characteristics of the Cluster Placement. The framework registers a series of performance metrics to be related to each execution, both (i) user-centric (runtime, application/kernel time, application-specific); (ii) system-centric (physical/virtual CPU and network utilization).

Mapping Characteristics to Performance: The nature of the workload is extracted by an equivalence to one of the Dwarfs [1], and at least one representative benchmark. The representative benchmarks are executed beforehand over several possible Cluster Placements (hundreds), and the relevant metrics are gathered, thereby creating a performance matrix.

For a given target application, a series of Cluster Placements (in our experience, at least 40) are proposed to create an initial profile for the application over virtualized environments. CCA enables us to find relationships between the datasets of the target and the representative application of the corresponding Dwarf. Within the space obtained through dimensionality reduction, we find linear regressions between performance and placement, and therefore we can predict performance for new placements using interpolations. For the Structured Grid Dwarf, we obtained accuracy higher than 90% in performance prediction, when at least 50 data points are known.

4   Summary

The methodology based on applications requirements can assists researchers to define what is the best to solve their set of scientific applications. The methodology enables to define representative evaluation tests, including a model to define a representative benchmark, when the real application could not be used. Also, the GF allows a decision-making based on the performances of a set of applications and architectures and its relative importance. We made a case study for bioinformatics applications, in which some steps are detailed and where the methodology proved to be useful and relevant [3].

The proposed methodology based on Cluster Placement and VESPA was helpful in understanding how latency effects can be minimized by carefully constructing virtual clusters. The relationship between performance and Cluster Placement appears non-linear and complex, but by using CCA we were able to find linear relationships between two sets of relationships, enabling reasonably accurate predictions. The accuracy seems to depends on the type of Dwarf, whereby applications with higher frequency of communication are more difficult to predict.

References

1. Asanovic, K., Bodik, R., Demmel, J., Keaveny, T., Keutzer, K., Kubiatowicz, J., Morgan, N., Patterson, D., Sen, K., Wawrzynek, J., Wessel, D., Yelick, K.: A view of the parallel computing landscape. Commun. ACM **52**(10), 56–67 (2009). http://doi.acm.org/10.1145/1562764.1562783
2. Ferro, M., Mury, A.R., Schulze, B.: Manual de metodologia de análise operacional de sistemas de computação científica distribuída de alto desempenho. Relatórios de Pesquisa e Desenvolvimento do LNCC 01/2015, Laboratório Nacional de Computação Científica, Petropolis (2015). www.lncc.br
3. Ferro, M., Nicolás, M.F., del Rosario, Q., Saji, G., Mury, A.R., Schulze, B.: Leveraging high performance computing for bioinformatics: a methodology that enables a reliable decision-making. In: 16th IEEE/ACM International Symposium on Cluster, Cloud and Grid Computing, CCGrid 2016, Cartagena, 16–19 May 2016, pp. 684–692. IEEE Computer Society (2016)
4. Mc Evoy, G., Porto, F., Schulze, B.: A representation model for virtual machine allocation. In: International Workshop on Clouds and (eScience) Applications Management - CloudAM 2012. 2012 IEEE/ACM Fifth International Conference on Utility and Cloud Computing (2012)
5. Wagner, D., Mylander, W., Sanders, T.: Naval Operations Analysis, 3rd edn. Naval Institute Press, Annapolis (1999)

Hard Faults and Soft-Errors: Possible Numerical Remedies in Linear Algebra Solvers

E. Agullo[1], S. Cools[2], L. Giraud[1(✉)], A. Moreau[1], P. Salas[3], W. Vanroose[2], E.F. Yetkin[4], and M. Zounon[5]

[1] Inria, Bordeaux, France
luc.giraud@inria.fr
[2] University of Antwerp, Antwerp, Belgium
[3] Sherbrooke University, Sherbrooke, Canada
[4] Istanbul Kemerburgaz University, Istanbul, Turkey
[5] The University of Manchester, Manchester, UK

Abstract. On future large-scale systems, the mean time between failures (MTBF) of the system is expected to decrease so that many faults could occur during the solution of large problems. Consequently, it becomes critical to design parallel numerical linear algebra kernels that can survive faults. In that framework, we investigate the relevance of approaches relying on numerical techniques, which might be combined with more classical techniques for real large-scale parallel implementations. Our main objective is to provide robust resilient schemes so that the solver may keep converging in the presence of the hard fault without restarting the calculation from scratch. For this purpose, we study interpolation-restart (IR) strategies. For a given numerical scheme, the IR strategies consist of extracting relevant information from available data after a fault. After data extraction, a well-selected part of the missing data is regenerated through interpolation strategies to constitute a meaningful input to restart the numerical algorithm. In this paper, we revisit a few state-of-the-art methods in numerical linear algebra in the light of our IR strategies. Through a few numerical experiments, we illustrate the respective robustness of the resulting resilient schemes with respect to the MTBF via qualitative illustrations.

Keywords: Numerical resiliency · Hard fault · Soft fault · Numerical linear algebra · Krylov linear solvers · Eigensolvers

1 Introduction

One of the current challenge in high performance computing (HPC) is to increase the level of concurrency by using the largest number of resources operated at lower energy consumption. The use of these parallel resources at large scale leads to a significant decrease of the mean time between faults (MTBF) of HPC systems. To cope with these unstable situations, parallel applications have to be resilient, *i.e.*, be able to compute a correct output despite the presence of faults.

© Springer International Publishing AG 2017
I. Dutra et al. (Eds.): VECPAR 2016, LNCS 10150, pp. 11–18, 2017.
DOI: 10.1007/978-3-319-61982-8_3

To guarantee fault tolerance, two classes of strategies are required. One for the fault detection and the other for fault correction. Faults such as computational node crashes are obvious to detect while silent faults may be challenging to detect. To cope with silent faults, a duplication strategy is commonly used for fault detection [14] by comparing the outputs, while triple modular redundancy (TMR) is used for fault detection and correction [25]. However, the additional computational resources required by such replication strategies may represent a severe penalty. Instead of replicating computational resources, studies [7,26] propose a time redundancy model for fault detection. It consists in repeating computation twice on the same resource. The advantage of time redundancy models is the flexibility at application level; software developers can indeed select only a set of critical instructions to protect. Recomputing only some instructions instead of the whole application lowers the time redundancy overhead [19]. In some numerical simulations, data naturally satisfy well defined mathematical properties. These properties can be efficiently exploited for fault detection through a periodical check of the numerical properties during computation [9].

Checkpoint/restart is the most studied fault recovery strategy in the context of HPC systems. The common checkpoint/restart scheme consists in periodically saving data onto a reliable storage device such as a remote disk. When a fault occurs, a roll back is performed to the point of the most recent and consistent checkpoint. According to the implemented checkpoint strategy, all processes may perform the periodical record simultaneously. It is called a coordinated checkpoint [10,21,24]. In parallel distributed environments, synchronizations due to coordination may significantly degrade application performance [11,18]. To avoid synchronization, uncoordinated checkpoint may be employed combined with message logging protocols [5,15]. Many mechanisms have been developed to lower the overhead of the checkpoint/restart strategy [17,22,27]. However, the additional usage of resources (such as memory, disk) that is required by checkpoint/restart schemes may be prohibitive, or the time to restore data might become larger than the MTBF [8].

Algorithm based fault tolerance (ABFT) is a class of approaches in which algorithms are adapted to encode extra data for fault tolerance at expected low cost [6,12,13,20]. The basic idea consists in maintaining consistency between extra encoded data and application data. The extra encoded data can be exploited for fault detection and for loss data recovery. ABFT strategies may be excellent candidates to ensure the resilience of an application; however they induce extra costs for computing and storing the data encoding even when no fault occurs.

In this paper, we present numerical resilient methods for linear algebra problems that are the innermost numerical kernels in many scientific and engineering applications and also often ones of the most time consuming parts. We consider iterative methods that are widely used in many engineering applications. In addition to having attractive computational features for efficient parallel implementations, iterative methods are potentially more resilient. After a "perturbation" induced by a fault, the computed iterate can still serve as an initial guess as long as the key data that define the problem to solve, that are the matrix and the right-hand side for linear system solver or simply the matrix for eignesolvers, are not corrupted. We exploit this natural resilience potential to design robust resilience iterative solvers which may still converge in the presence of successive and possibly numerous faults.

2 Interpolation Policies

In the context of parallel distributed computing, common faults are hardware node crashes. To cope with such node crashes often referred to as hard faults, an interpolation strategy has been introduced first for GMRES in [16] and extended and generalized to CG and GMRES in [3] where its theoretical properties are further studied. It consists in computing meaningful values for the lost entries of the current iterate through the solution of a relatively small linear system. The recovered iterate is then used as a new initial guess to restart GMRES. We name Linear Interpolation this class of methods and denote it LI. An alternative interpolation approach is based on a linear least squares solution. We name Least Square Interpolation this class of more robust but potentially more expensive methods and denote it LSI in the sequel. These LI and LSI numerical resilient strategies are called Interpolation-Restart (IR) strategies.

Assumption 1. *In our parallel computational context, all the vectors or matrices of dimension n are distributed by blocks of rows in the memory of the different computing nodes but scalars or low dimensional matrices are replicated.*

For the sake of simplicity of exposure, we first describe the interpolation policies in the context of linear solvers and extend them to eigensolvers. We consider the solution of sparse linear system

$$Ax = b,$$

where the matrix $A \in \mathbb{R}^{n \times n}$ is nonsingular, the right-hand side $b \in \mathbb{R}^n$ and the solution $x \in \mathbb{R}^n$. According to Assumption 1, the right-hand side b and the coefficient matrix A are distributed according to a block-row partition as well as all vectors of dimension n generated during the solve step whereas scalars or low dimensional matrices are replicated on all nodes. Let N be the number of partitions, such that each block-row is mapped to a computing node. For all p, $p \in [1, N]$, I_p denotes the set of row indices mapped to node p. With respect to this notation, node p stores the block-row $A_{I_p,:}$ and x_{I_p} as well as the entries of all the vectors involved in the solver associated with the corresponding row indices of this block-row. If the block A_{I_p,I_q} contains at least one nonzero entry, node p is referred to as neighbor of node q as communication will occur between those two nodes to perform a parallel matrix-vector product. By $J_p = \{\ell, a_{\ell,I_p} \neq 0\}$, we denote the set of row indices in the block-column $A_{:,I_p}$ that contain nonzero entries and $|J_p|$ denotes the cardinality of this set.

When a node crashes, all available data in its memory are lost. We consider the formalism proposed in [16] in the same computing framework, where data loss are classified into three categories: *computational environment, static* data and *dynamic* data. The computational environment is all data needed to perform the computation (code of the program, environment variables, ...). Static data are those that are setup during the initialization phase and that remain unchanged during the computation. They correspond to the input data to the problem and include in particular the coefficient matrix A, the right-hand side vector b. Dynamic data are all data whose value may change during the computation. The Krylov basis vectors (e.g., Arnoldi basis, descent directions, residual, ...) and the iterate are examples of dynamic data. In particular, we assume that when a fault occurs, the crashed node is replaced and the associated computational environment and static data are restored.

The LI strategy, first introduced in [16], consists in interpolating lost data by using data from surviving nodes. Let $x^{(f)}$ be the approximate solution when a fault occurs. After the fault, the entries of $x^{(f)}$ are known on all nodes except the failed one. The LI strategy computes a new approximate solution by solving a local linear system associated with the failed node. If the node p fails, $x^{(LI)}$ is computed via

$$\begin{cases} x_{I_q}^{(LI)} = x_{I_q}^{(f)} & \text{for } q \neq p, \\ x_{I_p}^{(LI)} = A_{I_p,I_p}^{-1}\left(b_{I_p} - \sum_{q \neq p} A_{I_p,I_q} x_{I_q}^{(f)}\right). \end{cases} \tag{1}$$

Alternatively, LSI, relies on a least squares solution. Assuming that the node p has crashed, x_{I_p} is interpolated as follows:

$$\begin{cases} x_{I_q}^{(LSI)} = x_{I_q}^{(f)} & \text{for } q \neq p, \\ x_{I_p}^{(LSI)} = \underset{x_{I_p}}{\mathrm{argmin}}\left\|\left(b - \sum_{q \neq p} A_{:,I_q} x_q^{(f)}\right) - A_{:,I_p} x_{I_p}\right\|. \end{cases} \tag{2}$$

Those ideas can be extended to the solution of the standard eigenproblem of the form:

$$Au = \lambda u,$$

where $A \in \mathbb{C}^{n \times n}$, with $u \neq 0$, $u \in \mathbb{C}^n$ and $\lambda \in \mathbb{C}$. The couple (λ, u) is called an eigenpair of A where the vector u is an eigenvector with the associated eigenvalue λ.

The IR strategies consist in interpolating lost data by using non-corrupted data. Let $u^{(f)}$ be an approximated eigenvector when a fault occurs. After the fault, the entries of $u^{(f)}$ are correct, except those stored in the failed node p. Assuming that in a parallel distributed environment, the current eigenvalue λ_f is naturally replicated in the memory of the different computing nodes, we present two strategies to compute a new approximate eigenvector. The first strategy, referred to as linear interpolation and denoted LI, consists in solving a local linear system associated with the submatrices A_{I_p,I_p} of the failed node. The second one relies on the solution of a least squares interpolation and is denoted LSI. Those two alternatives result from considering $(\lambda_f, u^{(f)})$ as an exact eigenpair.

We may have a block row viewpoint, which defines the LI variant. If node p fails, LI computes a new approximation of the eigenvector $u^{(LI)}$ as follows

$$\begin{cases} u_{I_q}^{(LI)} = u_{I_q}^{(f)} & \text{for } q \neq p, \\ u_{I_p}^{(LI)} = (A_{I_p,I_p} - \lambda \mathcal{I}_{I_p,I_p})^{-1}(-\sum_{q \neq q} A_{I_p,I_q} u_{I_q}^{(f)}). \end{cases}$$

Alternatively, we can have a block column point of view, which leads to the LSI variant that computes $u^{(LSI)}$ via

$$\begin{cases} u_{I_q}^{(LSI)} = u_{I_q}^{(f)} & \text{for } q \neq p, \\ u_{I_p}^{(LSI)} = \underset{u_{I_p}}{\operatorname{argmin}} \|(A_{:,I_p} - \lambda \mathcal{I}_{:,I_p}) u_{I_p} - \sum_{q \neq p}(A_{:,I_q} - \lambda \mathcal{I}_{:,I_q}) u_{I_q}^{(f)})\|. \end{cases}$$

Here, $\mathcal{I}_{n,n} \in \mathbb{C}^{n \times n}$ is the identity matrix and we furthermore assume that $(A_{I_p,I_p} - \lambda I_{I_p,I_p})$ is non singular and that $(A_{:,I_p} - \lambda \mathcal{I}_{:,I_p})$ has full rank.

3 Numerical Experiments

In order to study the numerical behavior of the IR strategies, and illustrate their possible robustness and weaknesses, we consider three additional executions in our numerical experiments. To distinguish between the interpolation quality effect and possible convergence delay introduced by the restart, we also report on what is referred to as the Enforced Restart (ER) execution. It consists in enforcing the solver to restart at iteration f.

The faulty parallel environment is simulated with the following procedure as a sequence of crash nodes occurring at certain dates (iteration numbers). The iterations at which faults occur are decided following a pseudo-random Weibull probability distribution, considered as a relevant and realistic probabilistic model for characterizing the behavior of large-scale computational platforms [23].

In Fig. 1, we investigate the robustness of the IR strategies when the rate of faults is varied while the amount of recovered entries remains the same after each fault, that is 0.2%. Those experiments are conduced with a GMRES(100) using the kim1 matrix. An expected general trend that can be observed on that example is: the larger the number of faults the slower the convergence. When only a few faults occur, the convergence penalty is not significant compared to the non-faulty case. For a large number of faults, the convergence is slowed down but continues to take place. For instance, for an expected accuracy of 10^{-7}, the number of iterations with 40 faults is twice the one without fault. More information and details on those resilient numerical techniques can be found in [3].

For the eigensolver, we illustrate the resilience of the IR strategies designed for Jacobi-Davidson. In these experiments, we seek the five ($nev = 5$) eigenpairs whose eigenvalues are the closest to zero. To facilitate the readability and the analysis of the convergence histories plotted in this section, we only report on

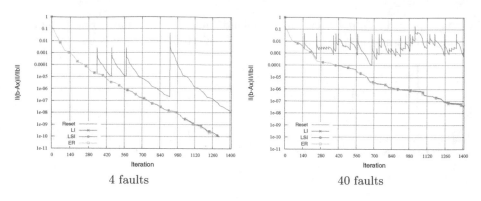

4 faults   40 faults

Fig. 1. Numerical behavior of the IR strategies when the rate of faults varies (matrix Kim/kim1 with 0.2% data loss at each fault)

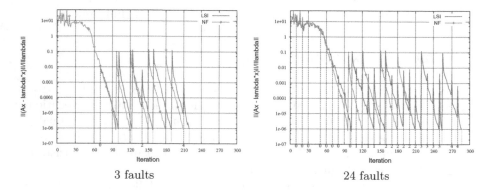

3 faults   24 faults

Fig. 2. Impact of the fault rate on the resilience of **LSI** when computing the **five eigenpairs** associated with the **smallest eigenvalues** using **Jacobi-Davidson**. The fault rate varies whereas a proportion of 0.2% of data is lost at each fault. (Color figure online)

results for one LSI strategy and the non faulty execution (NF). In these plots, we use vertical green lines to indicate the convergence of new eigenpairs (such as iterations 100, 122, 187 and 218 for the 3 fault graph in Fig. 2 and vertical red lines to indicate faulty iterations (such as iterations 70, 140 and 210 still in the 3 fault graph). We indicate the number of Schur vectors retrieved in the basis used to restart in red color under the vertical red line corresponding to the associated fault. For instance, 2 already converged Schur vectors are immediately retrieved at the restart, after the fault at iteration 140 in the 3 fault graph.

In Fig. 2 we depict the convergence histories when the fault rate varies leading to a number of faults that goes from 3 to 24; as expected, the larger the number of faults, the larger the convergence delay. However, the IR policy is rather robust and succeeds in converging the five eigenpairs in both cases. More information and details on those resilient numerical techniques for classical eigensolvers can be found in [4].

4   Concluding Remarks

Many scientific and engineering applications require the solution of linear algebra problems such as linear system solutions or eigenproblems. The objective of the work has been to propose and study numerical schemes suitable for the design of resilient parallel solvers. For that purpose, we have proposed two interpolation procedures to regenerate meaningful information for restarting the solvers after a fault. To evaluate the qualitative behavior of the resilient schemes, we have simulated stressful conditions by increasing the fault rate and the volume of data loss. One of the main features of this numerical remedy is that it does not require extra resources, *i.e.*, computational unit or computing time, when no fault occurs.

On the route of robust and resilient numerical linear algebra solvers, another challenge to address is related to soft-errors where the detection of a fault is already complex. This is an ongoing research activity we conduce and hope to have soon interesting results to present [1,2].

References

1. Agullo, E., Cools, S., Giraud, L., Vanroose, W., Yetkin, F.E.: On the sensitivity of CG to soft-errors and robust numerical detection mechanisms. Research Report in Preparation, Inria (2017)
2. Agullo, E., GiraudL, L., Moreau, A.: Adaptive soft-error detection criterion for GMRES. Research Report in Preparation, Inria (2017)
3. Agullo, E., Giraud, L., Guermouche, A., Roman, J., Zounon, M.: Numerical recovery strategies for parallel resilient Krylov linear solvers. Numer. Linear Algebra Appl. **23**, 888–905 (2016)
4. Agullo, E., Giraud, L., Salas, P., Zounon, M.: Interpolation-restart strategies for resilient eigensolvers. SIAM J. Sci. Comput. **38**(5), C560–C583 (2016)
5. Alvisi, L., Marzullo, K.: Message logging: pessimistic, optimistic, causal, and optimal. IEEE Trans. Softw. Eng. **24**(2), 149–159 (1998)
6. Anfinson, J., Luk, F.T.: A linear algebraic model of algorithm-based fault tolerance. IEEE Trans. Comput. **37**, 1599–1604 (1988)
7. Austin, T.M.: DIVA: a reliable substrate for deep submicron microarchitecture design. In: Proceedings of the 32nd Annual ACM/IEEE International Symposium on Microarchitecture, MICRO 32, Washington, DC, pp. 196–207. IEEE Computer Society (1999)
8. Cappello, F., Casanova, H., Robert, Y.: Preventive migration vs. preventive checkpointing for extreme scale supercomputers. Parallel Process. Lett. **21**, 111–132 (2011)
9. Chen, Z.: Online-ABFT: an online algorithm based fault tolerance scheme for soft error detection in iterative methods. In: ACM SIGPLAN Notices, vol. 48, pp. 167–176. ACM (2013)

10. Elnozahy, E.N., Alvisi, L., Wang, Y.-M., Johnson, D.B.: A survey of rollback-recovery protocols in message-passing systems. ACM Comput. Surv. **34**(3), 375–408 (2002)
11. Elnozahy, E.N., Johnson, D.B., Zwaenepoel, W.: The performance of consistent checkpointing. In: Proceedings of the 11th Symposium on Reliable Distributed Systems, pp. 39–47, October 1992
12. Gunnels, J.A., Van De Geijn, R.A., Katz, D.S., Quintana-ortí, E.S.: Fault-tolerant high-performance matrix multiplication: theory and practice. In: Dependable Systems and Networks, pp. 47–56 (2001)
13. Huang, K.-H., Abraham, J.A.: Algorithm-based fault tolerance for matrix operations. IEEE Trans. Comput. **33**, 518–528 (1984)
14. Iyer, R.K., Nakka, N.M., Kalbarczyk, Z.T., Mitra, S.: Recent advances and new avenues in hardware-level reliability support. IEEE Micro **25**(6), 18–29 (2005)
15. Johnson, D.B., Zwaenepoel, W.: Sender-based message logging (1987)
16. Langou, J., Chen, Z., Bosilca, G., Dongarra, J.: Recovery patterns for iterative methods in a parallel unstable environment. SIAM J. Sci. Comput. **30**, 102–116 (2007)
17. Li, C.-C.J., Fuchs, W.K.: Catch-compiler-assisted techniques for checkpointing. In: 20th International Symposium on Fault-Tolerant Computing. FTCS-20. Digest of Papers, pp. 74–81, June 1990
18. Liu, Y., Nassar, R., Leangsuksun, C.B., Naksinehaboon, N., Paun, M., Scott, S.L.: An optimal checkpoint/restart model for a large scale high performance computing system. In: IEEE International Symposium on Parallel and Distributed Processing (IPDPS 2008), pp. 1–9, April 2008
19. Oh, N., Shirvani, P.P., McCluskey, E.J.: Error detection by duplicated instructions in super-scalar processors. IEEE Trans. Reliab. **51**(1), 63–75 (2002)
20. Plank, J.S., Kim, Y., Dongarra, J.: Fault tolerant matrix operations for networks of workstations using diskless checkpointing. J. Parallel Distrib. Comput. **43**(2), 125–138 (1997)
21. Plank, J.: An overview of checkpointing in uniprocessor and distributed systems, focusing on implementation and performance. Technical report UT-CS-97-372, Department of Computer Science, University of Tennessee (1997)
22. Plank, J.S., Li, K.: ICKP: a consistent checkpointer for multicomputers. Parallel Distrib. Technol. Syst. Appl. **2**(2), 62–67 (1994). IEEE
23. Raju, N., Liu, Y., Leangsuksun, C.B., Nassar, R., Scott, S.: Reliability Analysis in HPC clusters. In: Proceedings of the High Availability and Performance Computing Workshop (2006)
24. Sancho, J.C., Petrini, F., Davis, K., Gioiosa, R., Jiang, S.: Current practice and a direction forward in checkpoint/restart implementations for fault tolerance. In: Proceedings of 19th IEEE International Parallel and Distributed Processing Symposium, April 2005
25. Scholzel, M.: Reduced triple modular redundancy for built-in self-repair in VLIW-processors. In: Signal Processing Algorithms, Architectures, Arrangements and Applications, pp. 21–26 (2007)
26. Vijaykumar, T.N., Pomeranz, I., Cheng, K.: Transient-fault recovery using simultaneous multithreading. In: Proceedings of the 29th Annual International Symposium on Computer Architecture, pp. 87–98 (2002)
27. Wang, C., Mueller, F., Engelmann, C., Scott, S.L.: Hybrid full/incremental checkpoint/restart for MPI jobs in HPC environments. Department of Computer Science, North Carolina State University (2009)

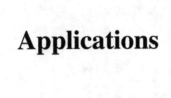

Applications

SIMD Parallel Sparse Matrix-Vector and Transposed-Matrix-Vector Multiplication in DD Precision

Toshiaki Hishinuma[1(✉)], Hidehiko Hasegawa[1,2], and Teruo Tanaka[2]

[1] University of Tsukuba, Tsukuba, Japan
hishinuma@slis.tsukuba.ac.jp
[2] Kogakuin University, Tokyo, Japan

Abstract. We accelerate a double-precision sparse matrix and DD vector multiplication (DD-SpMV) and its transposition and DD vector multiplication (DD-TSpMV) using SIMD AVX2. AVX2 requires changing the memory access pattern to allow four consecutive 64-bit elements to be read at once. In our previous research, DD-SpMV in CRS using AVX2 needed non-continuous memory load, processing for the remainder, and the summation of four elements in the AVX2 register. These factors degrade the performance of DD-SpMV. In this paper, we compare the storage formats of DD-SpMV and DD-TSpMV for AVX2 to eliminate the performance degradation factors in CRS. Our result indicates that BCRS4x1, whose block size fits the AVX2 register's length, is effective for DD-SpMV and DD-TSpMV.

Keywords: Matrix storage format · SpMV · Transposed SpMV · Double-double precision arithmetic · AVX2

1 Introduction

High-precision arithmetic operations reduce rounding errors and improve the convergence of Krylov subspace methods [1]; however, they are very costly. Double-double-precision (DD) arithmetic, which is one type of high-precision arithmetic, is constructed by combining double-precision operations, but it requires more than 10 double-precision operations for one DD operation [2]. However, it can greatly speed up performance using SIMD because it has a smaller memory access rate than double-precision arithmetic [4].

A sparse matrix and vector multiplication take much time in Krylov subspace methods. We accelerated the double-precision sparse matrix and the DD vector multiplication (DD-SpMV) and its transposition and the DD vector multiplication (DD-TSpMV) using advanced vector extensions 2 (AVX2) [3,4]. The AVX2 instruction set, which is a 256-bit single instruction multiple data streaming (SIMD) instruction set, provides fused multiply and add instruction (FMA). AVX2 simultaneously computes four double-precision FMA instructions.

© Springer International Publishing AG 2017
I. Dutra et al. (Eds.): VECPAR 2016, LNCS 10150, pp. 21–34, 2017.
DOI: 10.1007/978-3-319-61982-8_4

AVX2 required changing the memory access pattern to allow four consecutive 64-bit elements to be read at once. In DD-SpMV and DD-TSpMV for a compressed row storage format (CRS) [5], a non-continuous memory load and store are needed for using AVX2. In addition, since it must simultaneously compute four double-precision data. Furthermore, processing for the remainder in each row is needed, because AVX2 must simultaneously compute four double-precision data. Consequently, the performance might be degraded. We call these CRS problems, collectively, performance degradation factors.

To avoid them, we use the BCRS format [5], which divides matrix A into $r \times c$ small dense submatrices (called blocks), which might include some zero-elements. BCRS4x1, 2×2, and 1×4 ($r \times c = 4$) can simultaneously compute four elements.

BCRS4x1 ($r = 4$, $c = 1$) is suitable for DD-SpMV using AVX2 because the block size fits the SIMD register's length [6]. However, since BCRS4x1 requires up to four times the amount of operations and data as CRS. In DD-TSpMV, BCRS4x1 fails to eliminate performance degradation factors. Consequently, we must compare BCRS1x4 and BCRS4x1.

In this paper, we show that the effective implementation of DD-SpMV and DD-TSpMV improves the AVX2 performance and analyze the optimal storage format to eliminate the performance degradation factors in CRS.

2 Related Work

XBLAS [7] is a well-known extended precision BLAS whose input and output are double-precision that internally uses the DD operations. However, it does not accelerate them using SIMD.

Lis [8], which is an iterative solver library, internally uses DD operations, which are accelerated by SIMD SSE2. SSE2 has 128-bit SIMD registers.

On the other hand, Karakasis [9] and Im [10] accelerated a double-precision SpMV. Blocking, which fits the SIMD register's length, is effective for a double-precision SpMV. In AVX2, BCRS4x1 is effective. Xing [11] implemented a double-precision SpMV in ELLPACK and an ELLPACK sparse block format on MIC (Intel many integrated core architecture).

However, since these studies, which failed to evaluate TSpMV, are only double precision, we must compare DD-SpMV in BCRS1x4 and BCRS4x1.

3 Implementation of DD-SpMV and DD-TSpMV Using AVX2

3.1 DD Arithmetic

DD arithmetic, which is based on error-free, floating-point arithmetic algorithms by Dekker [12] and Knuth [13], only consists of combinations of double-precision values and uses two double-precision variables to implement one quadruple precision variable [2].

An IEEE 754 quadruple precision variable consists of a 1-bit sign part, a 15-bit exponent part, and a 112-bit significand part. A DD-precision variable consists of a 1-bit sign part, an 11-bit exponent part, and a 104-bit (52×2) significand part. The exponent part of a DD-precision variable is 4 bits shorter and the significand part is 8 bits shorter than the exponent and significand parts of an IEEE 754 quadruple precision variable, respectively.

The simplest way to use IEEE 754 quadruple precision is with Fortran REAL*16. We compared Fortran REAL*16 using an Intel Fortran compiler 13.0.1 (ifort) and DD arithmetic without any SIMD instructions. The compiler option in ifort was -O3. Fortran REAL*16 in ifort was only implemented by integer operations. We computed $y = \alpha \times x + y$, where x and y are the quadruple precision vectors and α is quadruple precision variable. Two 10^5 vectors, x and y, can be stored in the cache. The elapsed time of Fortran REAL*16 was 2.7 ms and that of the DD arithmetic was 0.64 ms in 1 thread, which means that the DD arithmetic was 4.2 times faster than Fortran REAL*16.

DD addition consists of 11 double-precision additions, and DD multiplication consists of 10 double-precision operations: three double-precision additions, three double-precision multiplications, and two double-precision FMA instructions ($3 + 3 + 2 \times 2 = 10$ flops).

In DD multiplication, two sign inversions are needed. However, since AVX2 lacks sign inversion instruction, we use two double-precision multiplications for two sign inversion. This flop count consists of these multiplications.

We implemented DD vector x using two double-precision arrays ($x.hi$ and $x.lo$) for the SIMD acceleration.

The bytes/flops of the DD operations are lower than those of the double-precision operations. For example, in the DD-SpMV kernel stored in CRS, the memory requirement is 28 bytes: 8 bytes for matrix A, 16-byte vector x, and 4 bytes for the vector column index. We postulate that loading vector x has a cache miss.

The bytes/flops of double-precision SpMV is 20 (bytes)/2 (flops) = 10, those of the DD matrix and the DD vector product is 36 (bytes)/23 (flops) = 1.56, and those of DD-SpMV is 28 (bytes)/21 (flops) = 1.33. The byte/flop value of DD-SpMV is 13% of double-precision SpMV and 85% of the DD matrix and the DD vector product.

DD-SpMV is expected to greatly speed up the SIMD acceleration because of the amount of data required for the memory. In many cases, for an iterative solver library, input matrix A, which is given in double precision, is iteratively used. To reduce the memory access of the sparse matrix and the vector product, we use double-precision sparse matrix A and DD-precision vector x product.

3.2   Intel SIMD AVX2

In this section, we introduce Intel SIMD AVX2. AVX2 must simultaneously compute, load, and store four double-precision variables. DD-SpMV and DD-TSpMV use three types of load AVX2 instructions (_mm256_load_pd (load), _mm256_broadcast_sd (broadcast), and _mm256_set_pd (set)) and one store instruction (_mm256_store_pd (store)).

These instructions have the following descriptions:

- The "load" instruction loads four continuous double-precision elements that begin with the same source memory address.
- The "broadcast" loads one double-precision element from one source memory address to all the elements of the SIMD register.
- The "set" loads four double-precision elements from four different source memory addresses.
- The "store" stores four continuous double-precision elements from the register to the memory beginning with the same source address.

We easily implemented the following three macro-functions (i.e. "SCATTER", "REDUCTION", "FRACTION_PROCESSING") to SIMD-ize DD-SpMV and DD-TSpMV.

To perform random store operation "scatter," we implemented a "SCATTER" macro-function using "store" and ordinary instructions. We also implemented "SCATTER" to store the double-precision temporary array using the "store" instruction and stored valuables using ordinary double-precision store instructions.

To store the summation of the elements in the SIMD register storage into one source address, we implemented a "REDUCTION" macro-function that computes a summation of four DD variables in two SIMD registers (high and low). We implemented "REDUCTION" using the "shuffle" instruction, which rearranges the double-precision elements. For example, the AVX2 register has $\{a, b, c, d\}$ elements. First, the "shuffle" instruction makes $\{b, a, d, c\}$ elements of AVX2 register from $\{a, b, c, d\}$. Second, we operate the DD addition using the AVX2 of these registers, and then $\{a + b, a + b, c + d, c + d\}$ AVX2 register is made. Finally, we operate the DD addition using ordinary instructions for the second and third elements. 'REDUCTION" consists of "shuffle" instructions and 11 (DD addition using AVX2) + 11 (DD addition using ordinary instruction) = 22 flops.

To judge the processing for the AVX2 remainder, which is one, two, or three elements for each row in the case of CRS, we implemented a "FRACTION_PROCESSING" macro-function, which assigns zero to the "set" operand at the execution and three conditional branchings.

"FRACTION_PROCESSING" consists of the following C code:

```
av = load(A[value[j]]);
yv = set_zero();
case (r == 3)
    xv2 = set(x[index[j]], x[index[j+1]], x[index[j+2]], 0);
    yv2 = DD_MULT_ADD(yv2, av, xv2); break;
case (r == 2)
    set(x[index[j]], x[index[j+1]], 0, 0);
    yv2 = DD_MULT_ADD(yv2, av, xv2); break;
case (r == 1)}
    set(x[index[j]], 0, 0, 0);
    yv2 = DD_MULT_ADD(yv2, av, xv2);
```

The av is the 256-bit AVX2 register type variables. $Xv2$ and $yv2$ are the AVX2 register type variables for DD variables. "DD_MULT_ADD" computes DD multiplication and addition using AVX2: $yv2 + av \times xv2$. "Set_zero()" means an AVX2 register type variable initialization with zero. "FRAC-TION_PROCESSING" needs a maximum of three times as many branches on the conditions.

The "set" is costly compared to "load" and "broadcast [3]." "SCATTER," "REDUCTION," and "FRACTION_PROCESSING" are costly because "SCAT-TER" occurs in the random memory store. "REDUCTION" requires more computations because it needs a "shuffle" instruction and 22 double-precision additions. "FRACTION_PROCESSING" occurs in the conditional branching.

4   Performance Degradation Factors of DD-SpMV and DD-TSpMV in CRS Using AVX2

4.1   DD-SpMV

The CRS format is expressed by the following three arrays: ind, ptr, and val. The double-precision val array stores the values of the non-zero elements of matrix A since they are traversed row-wise. The ind array is the column indices that correspond to the values, and ptr is the list of value indexes where each row starts. DD-SpMV in CRS using AVX2 consists of the following C code:

```
#pragma omp parallel for private (j, av, xv2, yv2)
for(i=0; i<N; i++){
    yv2 = set_zero();
    for(j=A >ptr[i]; j<A->ptr[i+1]-3; j+=4){
        xv2 = set(x[A->ind[j+0]],..,x[A->ind[j+3]]);
        av = load(&A->val[j]);
        yv2 = DD_MULT_ADD(yv2, av, xv2); break;
    }
    yv2 = FRACTION_PROCESSING();
    y[i] = REDUCTION(yv2);
}
```

X and y are a double-precision array, and A is the CRS format. DD-SpMV in CRS using AVX2 needs a "set" of x, the "REDUCTION" of y, and "FRAC-TION_PROCESSING."

4.2   DD-TSpMV in CRS Using AVX2

DD-TSpMV in CRS using AVX2 consists of the following C code:

```
num_threads = omp_num_threads();
work = malloc(num_threads * N);
#pragma omp parallel private (i, j, k, av, xv2, yv2){
```

```
   k = omp_get_thread_num();
#pragma omp for
   for(i=0; i<N; i++){
      xv2 = broadcast(&x[i]);
      for(j=A->ptr[i]; j<A->ptr[i+1]-3; j+=4){
         jj = j + k + N;
         yv2 = set(y[A->ind[j+0]],..,y[A->ind[j+3]]);
         av = load(&A->val[j]);
         yv2 = DD_MULT_ADD(yv2, av, xv2); break;
         SCATTER(yv2, work[A->ind[jj+0],..,work[A->ind[jj+3]]);
      }
      av = load(&A->val[j]);
      yv2 = FRACTION_PROCESSING(A,x);
      SCATTER(yv2, work[A->ind[jj+0],..,work[A->ind[jj+3]]);
   }}
for(i=0;i<N,i++)
   for(j=0;j<num_threads,i++)
      y[i] = DD_ADD(y[i], work[A->ind[i+j*N]]);
```

DD-TSpMV in CRS needs a "set" of y, a "SCATTER" of y, and "FRAC-TION_PROCESSING."

In multi-threading, DD-TSpMV in CRS needs the number of thread work vectors and their array-reduction after computation.

5   Implementation and Evaluation of DD-SpMV and DD-TSpMV in Other Storage Formats

CRS has some performance degradation factors, and AVX2 must change the memory access pattern to allow four consecutive 64-bit elements to be read at once. In this section, we compare the features of some storage formats using performance degradation factors.

5.1   DD-SpMV

BCRS $r \times c$ is expressed by the following three arrays: *bind*, *bptr*, and *bval*. The length of the double-precision array *bval* is the number of blocks (blk) $\times$ $r \times c$ store values of the non-zero blocks since they are traversed row-wise. The *bind* array is the column indices that correspond to the blocks, and *bptr* is the list of block indexes where each block row starts.

Table 1 shows the features of CRS, BCRS1x4, BCRS4x1, and ELL [5].

BCRS4x1 does not need "set," "REDUCTION," or "FRACTION_PROCESSING." It needs "REDUCTION," and ELL needs a "set" of x. In DD-SpMV, BCRS4x1 is the best estimation because it eliminates the performance degradation factors in CRS. However, it needs more operations and data.

Table 1. Features of DD-SpMV in each storage format

	CRS	BCRS1x4	BCRS4x1	ELL
Loading x	set	load	broadcast	set
Loading y	set_zero	set_zero	set_zero	set_zero
Storing y	REDUCTION	REDUCTION	store	store
FRACTION_PROCESSING	each row	none	none	each col.
Computation ratio (max)	1	4	4	the num. of row

DD-SpMV in BCRS4x1 using AVX2 consists of the following C code:

```
#pragma omp parallel for private (jb, av, xv2, yv2)
for(i=0; i<N-3; i+=4){ // block_row is about N/4.
   yv2 = set_zero();
   for(j=A->bptr[i]; j<A->bptr[i+1]; j++){
      xv2 = broadcast(x[A->bind[j]]);
      av = load(&A->bval[j * 4]);
      yv2 = DD_MULT_ADD(yv2, av, xv2); break;
   }
   y[i] = store(yv2);
}
```

In the inner-loop (j-loop), DD-SpMV in CRS needs four double-precision elements of A and four non-contiguous and indirect DD elements of x. BCRS4x1 only needs four double-precision elements of A and one indirect DD element of x. The amount of bytes/flops of BCRS4x1 is smaller than that of CRS. The memory requirement of BCRS4x1 in the inner-loop is smaller than that of CRS.

5.2 DD-TSpMV

The performance degradation factors of DD-TSpMV in CRS are non-continuous load/store, "FRACTION_PROCESSING," and the initialization and summation of the work vectors in multi-threading. In DD-SpMV, the BCRS4x1 feature is the best. However, in DD-TSpMV, BCRS4x1 fails to eliminate the work vectors.

We improved DD-TSpMV in BCRS4x1 for high performance in DD-SpMV and DD-TSpMV on only one storage format. Its BCRS4x1 applied column-wise multi-threading; the others applied row-wise multi-threading. The DD-TSpMV performance in BCRS4x1 applied additional column-wise multi-threading, which is improved here because BCRS4x1 only computes one column in j-loop; i.e., it can easily be thread-partitioned. DD-TSpMV in BCRS4x1 using the AVX2 of column-wise multi-threading consists of the following C code:

```
num_threads = omp_num_threads();
work = malloc(4* N);  // The length of SIMD.
#pragma omp parallel private(work, jb, av, xv2, yv2){
```

```
k = omp_get_thread_num();
alpha = N / num_threads * k;
beta = N / num_threads * (k+1);
for (i = 0; i < N-3; i+=4){
   xv2 = load(x[i]);
   #pragma omp for
   for (j = bptr[i]; j < bptr[i+1]; j++){
      if ( alpha < bind[jb] <= beta){ //thread-partitioning
      av = load(A->bval[j]);
      yv2 = broadcast(work[bind[j]]);
      yv2 = DD_MULT_ADD(yv2, av, xv2); break;
      y[i] = store(yv2);
}}}}
```

Table 2. DD-TSpMV features in each storage format

	CRS	BCRS1x4	BCRS4x1	ELL
Loading x	broadcast	broadcast	load	broadcast
Loading y	set	load	broadcast	set
Storing y	SCATTER	store	store	REDUCTION
Fraction_processing	each row	none	none	each col.
Computation ratio (max)	1	4	4	number of rows

BCRS4x1 can eliminate "REDUCTION" and continuously store work vectors in j-loop. Since it needs only four work vectors, it is expected to speed up performance on more multi-core systems.

Table 2 shows the TSpMV features in each storage format. BCRS1x4 and BCRS4x1 do not need "set" "scatter," or "REDUCTION." ELL needs "set" and "REDUCTION." In addition, BCRS4x1 only needs four work vectors and continuous storage for them. In DD-TSpMV, BCRS1x4 or BCRS4x1 is the best.

6   Experimental Results

We performed our tests on a machine with a 4-core 8-thread Intel Core i7 4770 3.4 GHz CPU, an 8-MB L3 cache, and 16-GB memory. We used Fedora 20 OS and Intel C/C++ compiler 15.0.0 as well as compiler options -O3, -xCORE-AVX2, -openmp, and -fp-model precise. Our code was written in C and used AVX2 intrinsic instructions. We also used an openMP guided scheduling option and 4-thread multi-threading.

6.1 DD-SpMV and DD-TSpMV Overheads

We evaluated the performance in each storage format with 23 matrices, which were taken from The University of Florida Sparse Matrix Collection (Florida Collection) [14].

Figures 1 and 2 shows the overhead of DD-SpMV and DD-TSpMV. We measured the elapsed time of the non-continuous load to change the load instruction from the set instruction. calculation kernel means elapsed time without overheads.

From Fig. 1, we compared the DD-SpMV in CRS overheads in the following results, where the calculation kernel is the baseline:

- Non-continuous load (set instruction) overheads are 74–630%.
- "FRACTION_PROCESSING" overheads are 4–89%.
- "REDUCTION" overheads are 27–380%.

The effect of the non-continuous load and "REDUCTION" is very large. When the nnz/row is small, the overhead effects are large because "FRACTION_PROCESSING" and "REDUCTION" occurred in each row.

The total time of BCRS1x4 is 1.6–7.3 times slower than the calculation kernel in CRS. The elapsed time of BCRS1x4 is more than four times slower because it needs "REDUCTION".

In all cases, BCRS1x4 is slower than BCRS4x1 because of the "REDUCTION" overhead. The total time of BCRS4x1 is 1.2–3.2 times slower than the

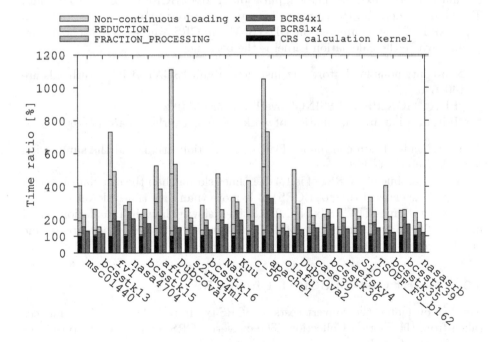

Fig. 1. DD-SpMV overhead

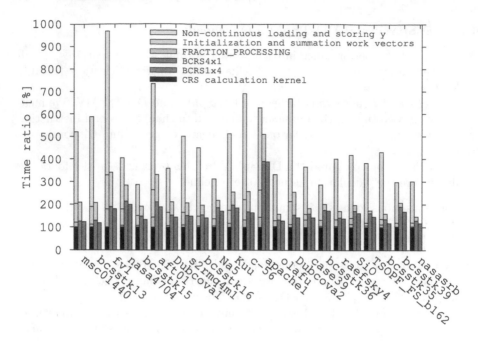

Fig. 2. DD-TSpMV overhead

calculation kernel in CRS. The computation ratios of BCRS4x1 are 1.1–3.9 times. The elapsed time and the computation ratio are proportional.

From Fig. 2, we compared the DD-TSpMV in CRS overheads in the following results, where the calculation kernel is the baseline:

- Non-continuous load/store (set instruction and "SCATTER") overheads are 140–640%.
- "FRACTION_PROCESSING" overheads are 3–79%.
- "Initialization and summation of work vectors" overheads are 12–150%.

The overheads of non-continuous load/store, Initialization, and the summation of work vectors are large.

The total time of BCRS1x4 is 1.4–5.1 times slower than the calculation kernel in CRS. The elapsed time of BCRS1x4 is more than four times slower.

The total time of BCRS4x1 is 1.2–3.8 times slower than the calculation kernel in CRS. The computation ratios of BCRS4x1 are 1.1–3.9 times. The elapsed time and the computation ratio are proportional.

6.2   Convert Costs of BCRS4x1 from CRS

Next we evaluated the convert costs of BCRS4x1 from CRS using 100 matrices taken from the Florida Collection. The convert BCRS4x1 from CRS consists of the following C code:

```
for(bi=0;bi<nr;bi++){
  i = bi*r;
  ii = 0;
  kk = Aout.bptr[bi];
  while(i+ii<n && ii<=r-1){
    for(k=Ain.ptr[i+ii];k<Ain.ptr[i+ii+1];k++){
      ......
          Aout.bindex[kk] = Ain.index[k]/c;
        Aout.value[ij] = Ain.value[k];
        kk = kk+1;
      ......
      }
    ii = ii+1;
    }
  ......
}
```

We measured the convert times and compared them to the computation times of DD-SpMV in BCRS4x1. The average convert time was 4.2, the minimum convert time was 2.6, and the maximum convert time was 5.1 times slower than the elapsed time of DD-SpMV in CRS. The convert time is small.

6.3 BCRS4x1 Effect

Figure 3 compares the time ratio of BCRS4x1 to CRS for 100 sparse matrices taken from the Florida Collection.

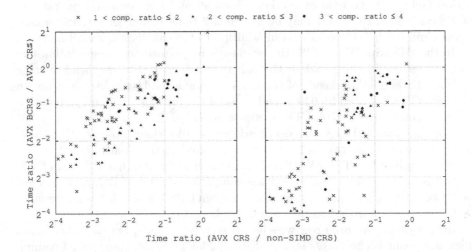

Fig. 3. Time ratio of BCRS4x1 compared with CRS [ms] (left: DD-SpMV, right: DD-TSpMV). The comp. ratio compares the amount of operations in BCRS4x1 to CRS.

In many cases, BCRS4x1 is faster than CRS using AVX2. The time ratios of DD-SpMV in CRS using AVX2 are 0.06–1.34 with an average of 0.38 compared with CRS without SIMD. The time ratios of DD-SpMV in BCRS4x1 are 0.09–1.96 times faster (average 0.43) than the case of CRS using AVX2. In DD-SpMV, the 96/100 matrix performance outperforms Scalar CRS.

The time ratios of DD-TSpMV in CRS using AVX2 are 0.06–1.14 with an average of 0.34 compared with CRS without SIMD. The times of DD-TSpMV in BCRS4x1 using AVX2 are 0.08–1.07 with an average of 0.38 compared with CRS using AVX2. In DD-TSpMV, the 100/100 matrix is better than Scalar CRS.

For example, the time ratio of "cell2" in BCRS4x1 using AVX2 is 1.96 times slower than that in CRS using AVX2. It has different placement of the non-zero elements in each row, nnz/row is 5, and the computation ratio is 1.9. When nnz/row is less than 8, BCRS4x1 is bad because it cannot improve the memory access.

7   Discussion

In DD-SpMV in CRS, the overheads by performance degradation factors are 130–1010% compared to the calculation kernel. The average overhead is about 300%. The overheads of the non-continuous load (set instruction) are 74–630%, and those of "REDUCTION" are 27–380%. These effects are very large.

BCRS4x1 may require at most four times the elapsed time of the calculation kernel. However, solving this trade-off problem is easy because CRS has 300% overheads.

As a result of DD-SpMV, BCRS4x1 is faster in 97/100 matrices than CRS. BCRS4x1 is effective for most cases because it eliminates the performance degradation factors. If the case of overhead is small and the computation ratio of BCRS4x1 is four, BCRS4x1 may be less than 2 times slower than CRS because it has a minimum 100% overhead in addition to the calculation kernel alone.

In the DD-TSpMV in CRS, the overheads by performance degradation factors are 180–890% compared to the calculation kernel. The overhead average is about 370%. The overheads of the non-continuous load/store (set instruction. and "SCATTER") are 140–640%, and those of the Initialization and summation of work vectors are 12–150%. These effects are huge.

The maximum DD-TSpMV overheads are smaller than those of DD-SpMV. However, since the average is large, any large-sized matrix is affected.

As a result of DD-TSpMV, BCRS4x1 is faster in 99/100 matrices than CRS. If the case of overhead is small and the computation ratio of BCRS4x1 is four, BCRS4x1 may be less than 1.4 times slower than CRS because it has a minimum 180% overhead compared to the calculation kernel.

In this paper, We used row-wise access storage formats. On the other hand, there are column-wise access storage formats, for example, compressed column storage (CCS) and Block CCS (BCCS).

In BCCS, since DD-TSpMV needs work vectors for multi-threading, the column-wise access storage formats are better in DD-TSpMV than DD-SpMV. In many algorithms, the frequency use of DD-SpMV exceeds DD-TSpMV. Row-wise access storage formats have higher versatility than column-wise access storage formats.

We conclude that the effects of eliminating the performance degradation factors are large. In sparse matrix operation with SIMD, we must eliminate non-continuous memory access and horizontal vector summation.

8   Conclusion

We evaluated the performance degradation factors of CRS using AVX2 and a storage format that eliminated the performance degradation factors of CRS for DD-SpMV and DD-TSpMV. We compared DD-SpMV and DD-TSpMV in CRS, BCRS1x4, and BCRS4x1 formats.

AVX2 required the memory access pattern to be changed to allow four consecutive 64-bit elements to be read at once. Four consecutive 64-bit elements must be allowed with blocking.

In DD-SpMV in CRS using AVX2, three performance degradation factors occur: non-continuous memory load from x, "FRACTION_PROCESSING," and the summation of the four DD variables in two SIMD registers. The overheads by the performance degradation factors are 130–1010% compared to the calculation kernel.

In DD-TSpMV in CRS using AVX2, three performance degradation factors occur: non-continuous load/store for y, the summation of each variable of SIMD register ("REDUCTION"), and "FRACTION_PROCESSING." However, DD-TSpMV in BCRS4x1 in multi-threading needs the number of thread work vectors and their summation.

One of our improvements is column-wise multi-threading, but such thread-partitioning is difficult for row-wise access storage format. Column-wise multi-threading of BCRS4x1, which can be easily implemented, can factor out the "REDUCTION" in the storage and summation of four work vectors.

In DD-TSpMV in CRS, the overheads by performance degradation factors are 180–890% compared to the calculation kernel.

BCRS4x1 may require at most four times the elapsed time of the calculation kernel. However, solving this trade-off problem is easy because the CRS overheads are large. If the overhead case is small and the computation ratio of BCRS4x1 is four, DD-SpMV in BCRS4x1 may be less than 2.0 times slower than CRS, and DD-TSpMV in BCRS4x1 may be less than 1.4 times slower than CRS. The convert cost of BCRS4x1 is about five times more than the computation time of DD-SpMV in BCRS4x1. This convert time is small.

BCRS4x1 is suitable for AVX2 because the block size fits the SIMD register's length and BCRS4x1 eliminates the performance degradation factors in CRS. However, BCRS4x1 requires at most four times the amount of operations and data as CRS. Changing the memory access pattern and thread-partitioning for the multi-threading are good implementation for DD-SpMV and DD-TSpMV.

In the future, we will apply our technique to other SIMD lengths and multi-core systems. Column-wise multi-threading in the BCRS format only needs the length of the SIMD's register work vectors because they are expected to speed up performance on multi-core systems.

Acknowledgments. This work was supported by JSPS KAKENHI Grant Number 25330144. The authors thank the reviewers for their helpful comments.

References

1. Kouya, T.: A highly efficient implementation of multiple precision sparse matrix-vector multiplication and its application to product-type Krylov subspace methods. Int. J. Numer. Methods Appl. **7**(2), 107–119 (2012)
2. Bailey, D.H.: High-precision floating-point arithmetic in scientific computation. Comput. Sci. Eng. **7**, 54–61 (2005)
3. Intel. http://software.intel.com/en-us/articles/intel-intrinsics-guide
4. Hishinuma, T., Fujii, A., Tanaka, T., Hasegawa, H.: AVX acceleration of DD arithmetic between a sparse matrix and vector. In: Wyrzykowski, R., Dongarra, J., Karczewski, K., Waśniewski, J. (eds.) PPAM 2013. LNCS, vol. 8384, pp. 622–631. Springer, Heidelberg (2014). doi:10.1007/978-3-642-55224-3_58
5. Barrett, R., et al.: Templates for the Solution of Linear Systems: Building Blocks for Iterative Methods, pp. 57–65. SIAM (1994)
6. Hishinuma, T., Fujii, A., Tanaka, T., Hasegawa, H.: AVX2 acceleration of double precision sparse matrix in BCRS format and DD vector product. IPSJ Trans. Adv. Comput. Syst. **7**(4), 25–33 (2014). (in a Japanese)
7. Li, X., et al.: Design, implementation and testing of extended and mixed precision BLAS. ACS Trans. Math. Softw. **28**(2), 152–205 (2002)
8. Lis: Library of Iterative Solvers for Linear Systems. http://www.ssisc.org/lis/
9. Karakasis, V., Goumas, G., Koziris, N.: Exploring the effect of block shapes on the performance of sparse kernels. In: 2009 IEEE International Symposium on Parallel & Distributed Processing, pp. 1–8 (2009)
10. Im, E., Yelick, K., Vuduc, R.: SPARSITY: optimization framework for sparse matrix kernels. Int. J. High Perform. Comput. Appl. **18**(1), 135–158 (2004)
11. Liu, X., Smelyanskiy, M., Chow, E., Dubey, P.: Efficient sparse matrix-vector multiplication on x86-based many-core processors. In: 27th International Conference on Supercomputing, pp. 273–282 (2013)
12. Dekker, T.: A floating-point technique for extending the available precision. Numerische Mathematik **18**, 224–242 (1971)
13. Knuth, D.E.: The Art of Computer Programming: Seminumerical Algorithms, vol. 2. Addison-Wesley, Reading (1969)
14. The University of Florida Sparse Matrix Collection. http://www.cise.uhl.edu/research/sparse/matrices/

Accelerating the Conjugate Gradient Algorithm with GPUs in CFD Simulations

Hartwig Anzt[1], Marc Baboulin[2], Jack Dongarra[1], Yvan Fournier[3],
Frank Hulsemann[3], Amal Khabou[2(✉)], and Yushan Wang[2]

[1] Innovative Computing Laboratory, University of Tennessee, Knoxville, USA
[2] Laboratoire de Recherche en Informatique, Université Paris-Sud, Orsay, France
khabou@lri.fr
[3] EDF R&D, Clamart, France

Abstract. This paper illustrates how GPU computing can be used to accelerate computational fluid dynamics (CFD) simulations. For sparse linear systems arising from finite volume discretization, we evaluate and optimize the performance of Conjugate Gradient (CG) routines designed for manycore accelerators and compare against an industrial CPU-based implementation. We also investigate how the recent advances in preconditioning, such as iterative Incomplete Cholesky (IC, as symmetric case of ILU) preconditioning, match the requirements for solving real world problems.

1 Introduction

A significant gap exists in-between the availability of open-source software libraries for sparse linear algebra computations on accelerators, and what is actually used in an industrial environment. An example is the *Code_Saturne* [5] package, a general purpose Computational Fluid Dynamics (CFD) software developed and used at Electricité de France (EDF). Among the main reasons behind this situation is the limited experience of how open-source packages, often coming from an academic environment, fit the demands of an industrial setting. Another concern is whether the accelerator hardware specifications, in particular the limited memory bandwidth of graphics processing units (GPUs), are suitable for real world applications. In this position paper, we address these two concerns by evaluating the performance of different implementations of the Conjugate Gradient (CG) method for two benchmarks with a finite volume origin. As the iterative solution process plays a key role in the simulation algorithm — it can account for up to 80% of the computational time in *Code_Saturne* — the performance improvements are quickly reflected in the overall runtime of the CFD simulation. Thus the main contribution of this work is to show that the use of GPU-enabled sparse linear algebra libraries in the framework of industrial applications allows for significant performance improvements with minimal implementation effort.

The rest of the paper is organized as follows. In Sect. 2 we provide some background about the industrial code, the software libraries, and the benchmarks that we consider. Section 3 reviews some strategies known to enhance

© Springer International Publishing AG 2017
I. Dutra et al. (Eds.): VECPAR 2016, LNCS 10150, pp. 35–43, 2017.
DOI: 10.1007/978-3-319-61982-8_5

the performance of iterative solvers on GPUs. This includes the optimization of the sparse matrix vector product (SpMV) which typically dominates the performance of Krylov solvers such as CG, and the use of kernel fusion for enhanced data locality. We also review some of the latest ideas on preconditioning techniques suitable for fine-grained hardware parallelism. In Sect. 4, we report some experimental results obtained using the different software packages to solve the CFD problems. We conclude in Sect. 5.

2   Problem Setting and Software Framework

Code_Saturne [5] is a general purpose Computational Fluid Dynamics (CFD) software package developed and used at Electricité de France (EDF). It is based on a co-located finite volume approach, using a fractional time step method. This allows for any type of polyhedral mesh, though best results are usually obtained with regular, hexahedral meshes. The flux discretization uses a 2-point scheme, with contributions due to mesh non-orthogonalities added at the right-hand side and solved through sub-iterations. The matrix graph is thus based on face to cell adjacencies, leading to very sparse matrices. For a scalar variable on a hexahedral mesh, we have 7 non-zero entries per row (6 face neighbors + 1 diagonal). For a tetrahedral mesh, this even goes down to 5 non-zero entries per row. The benchmark problems we consider in this paper originate from Code_Saturne. As the problems are all symmetric and positive definite, they can be solved efficiently with the CG iterative solver. Enhancing the CG with a Jacobi preconditioner (diagonal scaling) typically improves both convergence and performance. We note that, in *Code_Saturne*, parallelism is handled via MPI and OpenMP. However, in our evaluation, we limit the parallelism to OpenMP, as we are considering single node performance only.

The **CUsparse** [10] software library is a collection of routines for sparse linear algebra computations on NVIDIA GPUs. It provides the main building blocks, such as the sparse matrix vector product kernel, matrix conversion routines, and incomplete LU (ILU) preconditioning techniques. Some basic iterative solvers such as CG are also available. Developed by NVIDIA, this library typically achieves very good performance on NVIDIA architectures.

MAGMA [8] is an accelerator-focused linear algebra library developed at the University of Tennessee. It provides backends for NVIDIA GPUs, Intel's Xeon Phi manycore accelerators (MIC), and any OpenCL-compatible system such as AMD GPUs. In addition being well-known for the dense linear algebra routines, MAGMA also contains a large variety of solvers, preconditioners, and eigensolvers for sparse linear systems. Comprehensive support for NVIDIA GPUs is provided, some basic routines and functionalities are also available in OpenCL and for the Xeon Phi.

ViennaCL [11] is a free open-source linear algebra and solver library written in C++. The functionality provided by ViennaCL overlaps significantly with the functionality provided by MAGMA. However, ViennaCL provides a unified interface for three fully supported compute backends using CUDA (for NVIDIA

GPUs), OpenCL (for cross-vendor GPU-support), and OpenMP (for multi-core CPUs). Also, in contrast to MAGMA, the compute backends in ViennaCL can be switched at runtime.

In the experimental evaluation, we consider two benchmarks from the EDF application:

- The `bundle` problem is generated using one regular hexahedral mesh. The matrices are built from a matrix of size 16384 (average nnz per row is 7) for which we duplicate the initial mesh to obtain larger systems and study the scalability. Figure 1(left) shows the geometry of the domain in the numerical simulation.
- The `bora` problem originates from a mostly hexahedral mesh, but includes face subdivisions at non-conformal mesh joining interfaces. As a result, most matrix rows have 7 non-zero entries, but some rows have a higher number of non-zero entries. In this benchmark, we solve a linear system of size 10196476 and the geometry of the compute domain is shown in Fig. 1(right).

For both benchmarks, the `bundle` and `bora`, the linear system to be solved is extracted from the Laplacian operator within the pressure correction step, with a right-hand side corresponding to the flow initialization. The matrix structure does not change over time, but the coefficients may change whenever the fluid properties vary. For these test matrices, we assume constant temperature and constant fluid properties, which means that the matrix coefficients remain constant.

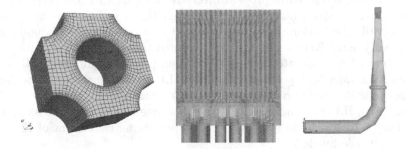

Fig. 1. Domain representation of `bundle` (left) and `bora` (right). The middle figure is a close-up of the upper part of the right figure. The discretization in this part is different from the rest of the domain, leading to a sparse matrix with an irregular pattern.

3 Sparse Linear Algebra on GPUs

The performance of Krylov methods like the Conjugate Gradient is generally bounded by the memory bandwidth of the hardware architecture used. Hence, optimizing the performance for these solvers is usually equivalent to optimizing the access to the GPU main memory. The implication is twofold: reducing the total amount of data that is read and written to the main memory, and organizing the memory access as coalescent reads [9]. As the CG and its precon-ditioned variant arise as a combination of matrix-vector and vector-operations, the optimization for coalescent memory reads boils down to the sparse matrix vector product. There exists extensive work on optimizing storage format and sparse matrix vector performance for GPUs, and in this work we focus on using the CSR, ELL, and SELLP formats, known to provide good performance [4]. To reduce the memory traffic, it is necessary to use algorithm-specific kernels that apply kernel-fusion to the basic linear algebra operations whenever possi-ble [1]. More precisely, consecutive vector operations sharing some of the input or output data are merged into a single kernel, such that data, once loaded into the fast multiprocessor memory, is reused. See [2] for details on how this is achieved for the Conjugate Gradient solver used in this study. The Magma library implementations feature kernel fusion for the basic CG as well as the preconditioned variant. In ViennaCL, the concept of kernel fusion is applied to the basic CG, not yet for the preconditioned variant. This optimization will be included in a future release, which will bring the performance of ViennaCL closer to that of MAGMA for the preconditioned CG as well. Beside Jacobi, another preconditioner suitable for a large variety of problems is an incomplete LU fac-torization [12]. A drawback of ILU preconditioners is the sequential nature of both the preconditioner generation via Gaussian-Elimination, and the sparse triangular solves in the preconditioner application. Also, approaches using level-scheduling or multi-color ordering for enhancing the concurrency often fail to exploit the fine grained parallelism provided by manycore architectures. Given this background, the recently proposed iterative approach to ILU precondition-ing has attracted much attention [6,7]. On GPUs in particular, the forward and backward substitutions traditionally used to solve the sparse triangular sys-tems in every outer Krylov iteration are expensive. Replacing those with a few Jacobi sweeps can accelerate the overall solution process significantly [3]. Vien-naCL and MAGMA both provide an iterative ILU, and we include this option in the experimental evaluation although, the *Code_Saturne* reference software does not contain an ILU preconditioner. In the experiments, as we are dealing with symmetric positive definite systems, we use the symmetric variant of ILU, the Incomplete Cholesky (IC).

4   Experimental Results

In this section, we analyze the convergence and performance of the Conjugate Gradient method using different preconditioners when solving the real-world CFD benchmarks previously described. The solvers are taken from different soft-ware libraries: *Code_Saturne* version 4.0.0 is compared against MAGMA release 2.0.0 and ViennaCL version 1.7.0. The GPU implementations are based on CUDA and CUsparse version 7.5 [10], and use an NVIDIA Tesla K40c GPU. The default block size is 256, which is also the size of the matrix slices in the SELLP format. *Code_Saturne* is using a 6-core Intel Xeon E5-2620 (Ivy Bridge) with hyperthreading enabled. In our experiments, we use 8 or 10 OpenMP threads, whichever provides the best performance.

In Fig. 2, we analyze how well the different CG implementations scale with respect to the problem size. As described in Sect. 2, we replicate the `bundle` problem to generate linear systems of larger dimensions. For a comprehensive evaluation, we use the MAGMA and ViennaCL solvers with different matrix storage formats. The intention is to identify the most suitable format for this problem. The right side shows the runtime of 100 iterations using a Jacobi-preconditioned CG (JCG). In this case, the preconditioner setup time is included as well. Note that CUsparse does not contain a pre-coded JCG implementation. ViennaCL only allows for the use of the CSR format, and does not provide a JCG version featuring kernel fusion in version 1.7.0. This explains the larger difference between the JCG runtime for ViennaCL and MAGMA when using the CSR format. As expected, SELLP again gives the best performance. A Jacobi preconditioner increases the pressure on the memory bandwidth, which is the performance-limiting factor for the GPU implementations. Nevertheless, the MAGMA JCG using SELLP format solves the largest problem about 8 times faster than *Code_Saturne*. For the small problems, the multicore JCG should be preferred. As SELLP gives also the best performance for the `bora` problem, we choose this format for the CG implementations of MAGMA and ViennaCL. On the left side, we show the time needed to execute 100 CG iterations using different combinations of software and matrix formats. For the GPU-based solvers (CUsparse, ViennaCL, MAGMA), the time needed for transferring the matrix and vectors between host and GPU is also included. 100 iterations are typically insufficient for convergence (also for this problem), which emphasizes the impact of these data transfers.

For small problems, the overhead of the data transfers plays an important role. Also, the parallel compute power of the GPU cannot be exploited, as the size of the linear system is smaller than the parallelism provided. For these

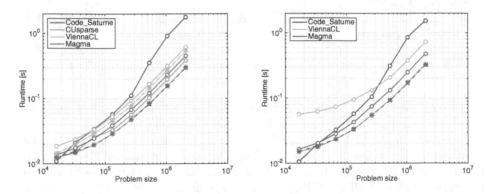

Fig. 2. Solver execution time for 100 iterations of different implementations of CG (left) and JCG (right). The target problem is the replicated `bundle`. Solid lines (with circle marker) are for CSR format, dotted lines (with star marker) for SELLP.

problems, *Code_Saturne* is much faster than all GPU codes. With increasing problem size, the runtime of *Code_Saturne* grows much faster than for the GPU implementations, and for the largest problem (2 million unknowns), ViennaCL, CUsparse, and MAGMA run between 5 and 10 times faster than the multicore CG. NVIDIA's CUsparse implementation is highly optimized, and its performance for the CSR format is unmatched by either MAGMA or ViennaCL. At the same time, it does not support the ELL and the SELLP matrix format, which gives much better performance for this class of test matrices. For MAGMA and ViennaCL, using SELLP gives the fastest CG execution time. The higher back-end flexibility of ViennaCL comes along with some performance decrease. The MAGMA implementation of CG is optimized in CUDA, and using the SELLP format in this routine is the overall winner for larger problems (15000 unknowns).

In Fig. 3, we compare the runtime for the different software libraries and solver settings when solving the **bora** problem. Notice that, in contrast to Fig. 2, we do not show the execution time for a fixed number of iterations, but show the timings of the preconditioner setup phase, the data transfer, and the iteration phase when solving the linear system for a relative residual stopping criterion of 10^{-10}. This implies that using a Jacobi preconditioner improves convergence, but makes every CG iteration more expensive. The validity of the results is ensured as the iteration counts are consistent across the different software libraries.

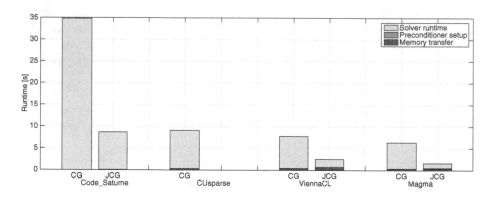

Fig. 3. Execution time of different implementations of CG and JCG for **bora**.

Despite the additional cost of the preconditioner setup, all implementations benefit from using a Jacobi preconditioner. The execution time for *Code_Saturne* improves by a factor of 4. Similarly, the ViennaCL and MAGMA JCG solve **bora** significantly faster than the corresponding CG implementations. The acceleration is smaller for ViennaCL, as the JCG is not enhanced with kernel fusion in the current release. All the GPU implementations are significantly faster than the multicore implementation. The overall winners are the MAGMA implementations for CG and JCG. Compared to the multicore *Code_Saturne* implementation, MAGMA improves the execution times of CG and JCG from 34.81 s to

6.36 s and from 8.64 s to 1.61 s, respectively. This includes the expensive pre-conditioner setup phase. In a scenario where a sequence of similar problems has to be solved, the reuse of a generated preconditioner would provide even larger benefits.

Although not available in *Code_Saturne*, we want to investigate whether the recent advances in iterative ILU preconditioning are suitable for the given real-world problems. ILU preconditioners are well-known to significantly reduce the iteration count for a large range of problems, and replacing the exact sparse triangular solves in the preconditioner application with approximate triangular solves can also make incomplete factorization preconditioners attractive for GPUs [3].

In Fig. 4, we compare iteration count (left) and execution time (right) for solving the **bora** problem with different preconditioners. Although also available in ViennaCL, in this experiment we focus on the MAGMA software package, as we exclusively target NVIDIA GPUs, and previously identified the implementations in MAGMA as performance winners for this problem. As previously mentioned, despite the ILU-notation, we internally use an incomplete Cholesky factorization, the symmetric variant of the incomplete LU factorization [12].

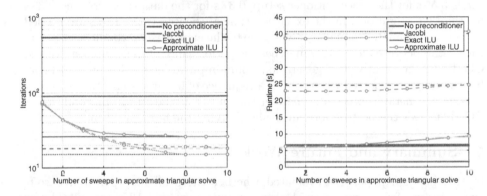

Fig. 4. CG iteration count (left) and execution time (right) for solving **bora** with different preconditioner configurations available in the MAGMA library. For the ILU preconditioner, solid lines are ILU(0), dashed lines are ILU(1), dotted lines are ILU(2). Exact ILU uses exact triangular solves based on level scheduling, approximate ILU uses different numbers of Jacobi sweeps.

The left side shows the iteration count needed for the relative residual stopping criterion of 10^{-10} when using a plain CG, a Jacobi preconditioner (JCG), an exact ILU preconditioner, or the variant using approximate triangular solves. The incomplete factorization preconditioners are generated as level-ILU [12] using different fill-in levels. For simplicity, the factorizations themselves are generated as exact factorizations, despite MAGMA also providing the functionality for iterative ILU-factor generation [6].

Using a Jacobi preconditioner decreases the number of iterations significantly, from 547 to 90. An exact factorization preconditioner provides even larger convergence improvement, reducing the iteration count to 26, 18, and 15, for ILU(0), ILU(1), and ILU(2), respectively. Using approximate triangular solves requires some additional iterations of the outer CG solver, but depending on the fill-in level, 3-6 sweeps in the approximate triangular solves are sufficient to bring the CG iteration count close to the exact ILU.

More relevant than the iteration count is the execution time, as this is the metric of interest when optimizing CFD simulations. The right side of Fig. 4 shows the corresponding execution time of the different configurations. Despite the higher iteration count, approximate triangular solves accelerate the ILU-preconditioned CG. Also, the lower iteration count for higher fill-in levels is not reflected in execution time, and despite the significantly lower iteration count (15 vs. 547), the ILU(2) using exact triangular solves needs about 8 times longer than the unpreconditioned CG. This comes partly from the higher cost of the preconditioner setup and data transfers. Also, higher fill-in levels make the sparse triangular solves (exact and approximate) more expensive. For using an incomplete factorization preconditioner, the runtime winner is the setting of an ILU(0) and two Jacobi sweeps in the approximate triangular solves. This configuration needs 3.55 s for the preconditioner setup, 0.37 s for the data transfers, and 2.38 s for the PCG iterations. In the execution time of the iterations, 1.97 s are needed for the approximate triangular solves. Due to the expensive preconditioner setup, the overall performance hardly matches the performance of the unpreconditioned CG. Using the iterative ILU generation would improve the results, but real benefits can only be expected when solving a sequence of linear systems that allow for reusing a generated preconditioner. In the end, it is the Jacobi-preconditioned CG that gives the best performance when solving the `bora` problem.

5   Summary and Future Work

In this position paper, we evaluated whether the available open-source software libraries for sparse linear algebra computations on GPUs are suitable for real-world problems arising from an industrial application. For two CFD simulations, we compared the performance of the solvers from *Code_Saturne*, an in-house developed multicore computational fluid dynamics code from EDF, to that of CUsparse, MAGMA and ViennaCL. The results reveal the superiority of *Code_Saturne* for small problems. For large problems, the GPU codes run up to 5× faster. In the future, we will address sequences of linear systems and evaluate the benefits of reusing a preconditioner for problems with similar properties. We will also address non-symmetric problems, where the benefits of the Jacobi preconditioner are typically smaller, and the iterative ILU preconditioner may become the method of choice.

Acknowledgements. This work was funded by the contract P02220 between Université Paris-Sud and EDF. We are grateful to Karl Rupp (TU Wien) for his support in using the ViennaCL library.

References

1. Aliaga, J.I., Pérez, J., Quintana-Ortí, E.S.: Systematic fusion of CUDA kernels for iterative sparse linear system solvers. In: Träff, J.L., Hunold, S., Versaci, F. (eds.) Euro-Par 2015. LNCS, vol. 9233, pp. 675–686. Springer, Heidelberg (2015). doi:10.1007/978-3-662-48096-0_52
2. Aliaga, J.I., Perez, J., Quintana-Orti, E.S., Anzt, H.: Reformulated conjugate gradient for the energy-aware solution of linear systems on GPUs. In: 2013 42nd International Conference on Parallel Processing (ICPP), pp. 320–329, October 2013
3. Anzt, H., Chow, E., Dongarra, J.: Iterative sparse triangular solves for preconditioning. In: Träff, J.L., Hunold, S., Versaci, F. (eds.) Euro-Par 2015. LNCS, vol. 9233, pp. 650–661. Springer, Heidelberg (2015). doi:10.1007/978-3-662-48096-0_50
4. Anzt, H., Tomov, S., Dongarra, J.: Energy efficiency and performance frontiers for sparse computations on GPU supercomputers. In: Proceedings of the Sixth International Workshop on Programming Models and Applications for Multicores and Manycores, PMAM 2015, pp. 1–10. ACM, New York (2015)
5. Archambeau, F., Méchitoua, N., Sakiz, M.: Code Saturne: A Finite Volume Code for the computation of turbulent incompressible flows - Industrial Applications. Int. J. Finite 1(1) (2004)
6. Chow, E., Anzt, H., Dongarra, J.: Asynchronous iterative algorithm for computing incomplete factorizations on GPUs. In: Kunkel, J.M., Ludwig, T. (eds.) ISC High Performance 2015. LNCS, vol. 9137, pp. 1–16. Springer, Cham (2015). doi:10.1007/978-3-319-20119-1_1
7. Chow, E., Patel, A.: Fine-grained parallel incomplete LU factorization. SIAM J. Sci. Comput. 37, C169–C193 (2015)
8. MAGMA Web page. http://icl.cs.utk.edu/magma/index.html
9. NVIDIA Corporation. CUDA C best practices guide. http://docs.nvidia.com/cuda/cuda-c-best-practices-guide/
10. NVIDIA Corporation. CUDA Toolkit Documentation v7.5, September 2015
11. Rupp, K., Rudolf, F., Weinbub, J.: ViennaCL - a high level linear algebra library for GPUs and multi-core CPUs. In: International Workshop on GPUs and Scientific Applications, pp. 51–56 (2010)
12. Saad, Y.: Iterative Methods for Sparse Linear Systems. Society for Industrial and Applied Mathematics, Philadelphia (2003)

Parallelisation of MACOPA, A Multi-physics Asynchronous Solver

Ronan Guivarch[1]([✉]), Guillaume Joslin[1], Ronan Perrussel[2], Daniel Ruiz[1],
Jean Tshimanga[1], and Thomas Unfer[2]

[1] INP(ENSEEIHT)-IRIT, University of Toulouse, Toulouse, France
`ronan.guivarch@enseeiht.fr`
[2] INP(ENSEEIHT)-LAPLACE, University of Toulouse, Toulouse, France

Abstract. Macopa is a partial differential equations solver based on a particular local time-stepping technique dedicated to multi-physics and multi-scale problems. Here, some parallelisation strategies – multithreading, domain decomposition, and hybrid OpenMP/MPI – are introduced for this solver. Their efficiency is evaluated on a few examples.

1 Context

Numerical simulation has become a central tool for the modeling of many physical systems (combustion, atmospheric plasmas, etc.). Multi-scale phenomena make the integration of these models difficult in terms of accuracy and computation time. Time-stepping integration techniques used for modeling such problems generally fall into two categories: explicit and implicit schemes. In the explicit schemes, all unknown variables are computed at the current time level from quantities already available. Time step is then limited by the most restrictive stability condition over the whole computation domain. In the implicit method, the time step is no longer limited by a stability condition. However the scheme is generally not suitable for strongly coupled problems. To solve such problems, a number of local time-stepping approaches have been developed. These methods are restricted by a local stability condition rather than the traditional global stability condition.

Macopa is a Partial Differential Equations (PDE) solver based on an asynchronous time-stepping technique proposed in [1]. The asynchronous time-stepping is an explicit local time-stepping technique which is consistent in time for solving a system of conservative PDE. Two data description modes are considered, cell-centered schemes and cell-vertex schemes. It has been successfully applied to fluid mechanics, combustion, micro-wave propagation, and plasma discharge modeling. Recent developments have extended the paradigm to higher order accuracy when it is used in combination with a discontinuous Galerkin method [2].

The capability of handling a large number of different time steps is obtained assuming that the time steps themselves can be discretized using an elementary virtual sub-time step. Under this hypothesis a Discrete Time Scheduler (DTS) has been introduced in [1]. The architecture of the DTS relies on two concepts:

© Springer International Publishing AG 2017
I. Dutra et al. (Eds.): VECPAR 2016, LNCS 10150, pp. 44–51, 2017.
DOI: 10.1007/978-3-319-61982-8_6

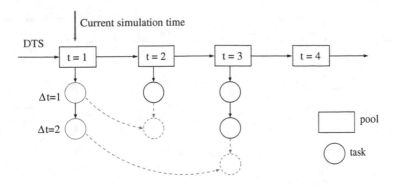

Fig. 1. Example of a sequential DTS

the "task" which is an elementary computation to be done, typically "refreshing" the values of the local variables in a cell of the mesh, and the "pool" which is a list of tasks to be treated at a given time tag. This list can eventually be empty when no task are required at this given time tag. The DTS itself is a circular table of pools with a particular pool holding the current time. As the simulation moves in time, a new time tag in the future is assigned to the previous pool (see Fig. 1). The horizon of the DTS is the size of the sub-time step multiplied by the number of pools in the table. Tasks, which have time steps larger than the horizon of the DTS, are managed with delays. The DTS insures a planning of n tasks with a complexity of $\mathcal{O}(n)$. When it is required, the virtual sub-time step is rescaled dynamically during the simulation.

2   Parallelisation

Parallelisation of the asynchronous time-stepping was realized using either multi-threading or mesh partitioning with possible hybridization of both approaches.

2.1   Multi-threading

Multi-threading of the algorithm is possible as long as thread-safety is insured when accessing the data within the mesh, but it also has to be handled when performing the scheduling of the tasks.

For thread-safe scheduling of the tasks, the strategy that has been followed is to duplicate the DTS: creating one circular table of pools for each thread. At the beginning of the treatment of the current simulation time, task lists are balanced from one local DTS to the others if needed (see Fig. 2). Then each thread processes its tasks and replaces them in the future pools within its local DTS.

The mesh data issues arises because for conservation equations, say for instance in the finite volume approach, data from the cells on both side of a face are needed to compute a numerical flux through the face. Then this flux is

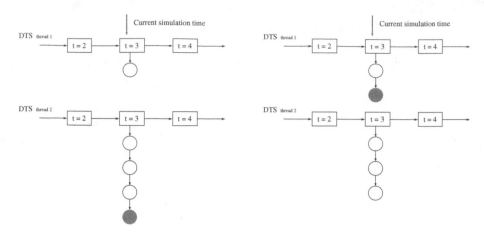

Fig. 2. Example of a multi-threaded DTS: before starting the treatment at the current simulation time (left), the Master balances the tasks between the threads (right).

used in the flux balance computation needed to obtain the time derivatives of the unknown in both neighboring cells. So if two threads try to refresh two adjacent cells at the same time, a thread-safety issue occurs. The workaround consists in splitting the algorithm in two phases. During the first phase, threads treat their tasks. They are allowed to access cell data such as states and time derivatives for reading but not for writing. In this phase, only intermediate variables such as fluxes or source terms are written, and for each cell in which the states and time derivative have to evolve an integration task is then created. This task is inserted into the pool of the thread creating it (so that those updates can be done safely in parallel). The integration task consists in synchronizing the states at present simulation time and computing the new time derivatives. For the integration to be thread safe, we must insure that a single cell appears only once in the integration lists of all threads. A synchronization step is performed to remove every duplicated cell from the union of these lists of integration tasks. Then the size of the lists are balanced again and each thread performs its integration tasks.

2.2   Mesh Partitioning

Mesh partitioning has been done using the SCOTCH mesh partitioner [3] (Fig. 4). The mesh cells have been weighted with the refreshment frequency (inverse of the local time step) and the faces are weighted with the average refreshment frequency of both cells. Process communications are done using MPI. Each process has its own DTS. For two adjacent cells on both side of the MPI boundary that have different time steps, the refreshment time tags do not necessarily match. The approach that has been developed is to send fluxes/partial residuals with their time of validity. The time of validity is the next simulation time tag at which the MPI boundary cell must be updated again.

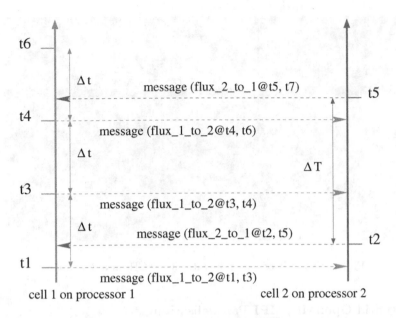

Fig. 3. Example of communication between two processors: each message contains, for two adjacent cells, some physical fields and the time of validity of those fields. (Color figure online)

So from the sender point of view, a message shall be sent each time an intermediate variable (flux or partial residual) is recomputed. The receiver has to receive as many messages as needed until the message that contains a validity time that is beyond its own simulation time. So that the current values can be incorporated into the current computations.

For instance, in Fig. 3, the blue processor 2 has to wait the message (flux_1_to_2@t4, t6) in order to perform its task at time t_5.

A single MPI message could be sent/received for every single cell at the MPI border, using for instance different message tags. But this approach faces MPI latency because of too many very small messages. The workaround that has been implemented is to manually create larger messages with all single messages that shall be sent/received at the current simulation time of a MPI process. All MPI calls are non blocking, so if a task needs a MPI message which is not available yet, it is pushed back at the end of the pool. By doing this, other tasks, which do not need MPI messages (such as interior cells), can be performed in the meantime.

Furthermore MPI cell tasks are considered more urgent than inner cell tasks, so when scheduling the task, MPI cell tasks are inserted on top of the pool whereas inner cell tasks are inserted at the bottom of the pool.

Note: with this paradigm, from the simulation time point of view, every process is also "asynchronous".

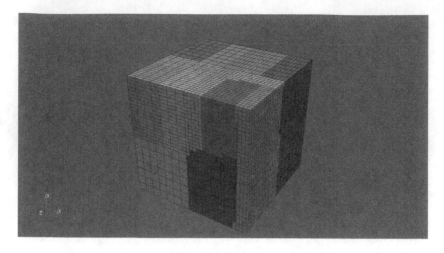

Fig. 4. Example of partition with SCOTCH on a 3D case

2.3 Hybrid OpenMP/MPI Parallelisation

For the hybrid mode, MPI is used in multiple thread mode so that every thread in every process can communicate via MPI to any other thread in a locally connected process. When solving PDE using the cell-vertex data description, two threads refreshing two adjacent cells could try to read the same boundary point MPI message. To insure thread-safety in this case, the OpenMP lock concept is used. A thread shall lock a boundary point prior to try to access to its MPI message.

3   Performance Results

We present in this section some performance results obtained on the super-computer EOS of CALMIP.

Its system is a *Bullx DLC B710 Blades, Intel Xeon E5-2680v2 10C 2.8GHz, Infiniband FDR* system. EOS got 122,440 cores and its performances are 255,078 Gflop/s as RMAX and 274,176 Gflop/s as RPEAK and is 399 in the 46th Top500 list (November 2015).

3.1 OpenMP Results

Our first test case is a 2D CH4/air premixed laminar flame (see Fig. 5 for a snapshot of the temperature field). The mesh consists of unstructured triangles for a total of 73, 850 nodes.

In Fig. 6 it is shown the speed-up obtained on 1 node where we vary the number of threads from 1 to 20 (number of cores by node). We notice that the results are perfectly scalable until 8 threads where the number of tasks by pool is probably the limiting factor.

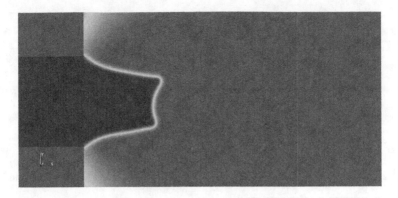

Fig. 5. Temperature field

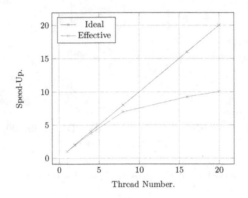

Fig. 6. Flame: OpenMP strong scaling

3.2 Domain Decomposition Results

Our test case for those results is the propagation of an acoustic wave. It is possible to simulate in a 2D or 3D simple domain.

For the 2D results, two meshes are used. The first mesh is uniform and made of quadrangles. In the second mesh, quadrangle size varies according to a polynomial law in both directions. Thus for the last one, the simulation faces a large amount of different local time steps.

The uniform mesh permits to validate the MPI strategy presented in Sect. 2.2. The non uniform mesh should illustrate the benefit of the asynchronous behavior of Macopa in a strongly asynchronous case.

Figure 7 shows that it is possible to implement a parallel version using MPI of the asynchronous algorithm with a reasonable speed-up. However our implementation is a first step and many optimizations are still possible (see the last section for some perspectives).

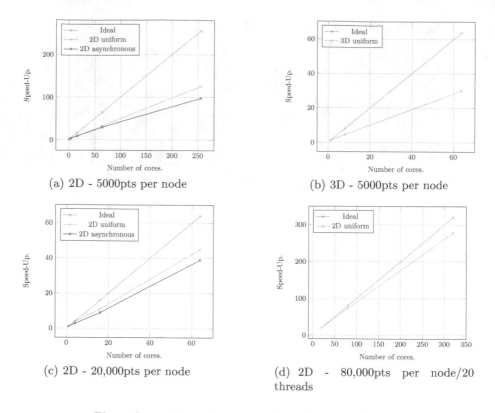

(a) 2D - 5000pts per node

(b) 3D - 5000pts per node

(c) 2D - 20,000pts per node

(d) 2D - 80,000pts per node/20 threads

Fig. 7. Speed-Up on the propagation of an acoustic wave

The results of Fig. 7(a) show that for the current parallel implementation of Macopa, 5,000 points per core are too low. The MPI boundaries are large with respect to the interior mesh. In this case, overlapping computation with communication is not effective. The results for the 3D case (Fig. 7(b)) shows the same trends. The performance is improved with more points per core (Fig. 7(c)).

The difference between the uniform and the asynchronous cases means there is still some improvements to propose for the mesh partitioning.

For the hybrid case (OpenMP + MPI), we assume that the speed-up with one node and 20 threads is linear. We know that this assumption is far too optimistic (see Sect. 3.1) but the point was to assess the MPI performance in the natural way of using the EOS super-computer: OpenMP on one node and MPI between the nodes. In this situation, with 80,000 points per node, the scaling is interesting (Fig. 7(d)).

4   Future Work

We noticed that it is difficult with SCOTCH to take into account the amount of work on each cell with asynchronous meshes in order to generate a balanced partitioning. We are looking for some other partitioners, for instance hypergraph partitioners [4, 5] in order to express better the constraints and obtain a better partitioning.

Finally, a new version of the standard MPI is now complete. For instance, the MPI Remote Memory Access (RMA) interface has been re-examined and permits efficient one-sided programming model within MPI. We should investigate these new functionalities to determine if they could be useful in Macopa.

Acknowledgements. This research is granted by the project MACOPA (ANR-11-MONU-0019).

This work was performed using HPC resources from CALMIP (Grants 2015-[p1528] and 2016-[p16023]).

We also thank Alfredo Buttari for his support and advices.

References

1. Unfer, T., Boeuf, J.P., Rogier, F., Thivet, F.: An asynchronous scheme with local time-stepping for multi-scale transport problems: application to gas discharges. J. Comp. Phys. **227**, 898–918 (2007)
2. Toumi, A., Dufour, G., Perrusel, R., Unfer, T.: Asynchronous numerical scheme for modeling hyperbolic systems. Comptes Rendus Mathematique **353**(9), 843–847 (2015)
3. Pellegrini, F.: SCOTCH 5.1 User's Guide, Laboratoire Bordelais de Recherche en Informatique (LaBRI) (2008)
4. Çatalyürek, Ü.V., Aykanat, C.: PaToH: A Multilevel Hypergraph Partitioning Tool, Version 3.0, Bilkent University, Department of Computer Engineering, Ankara, 06533 Turkey (1999). PaToH is available at http://bmi.osu.edu/umit/software.htm
5. Rietmann, M., Peter, D., Schenk, O., Uçar, B., Grote, M.J.: Load-balanced local time-stepping for large-scale wave propagation, IEEE CPS. In: 29th IEEE International Parallel & Distributed Processing Symposium, Hyderabad, India, pp. 925–935, May 2015. <hal-01159687>

Performance Analysis of SA-AMG Method by Setting Extracted Near-Kernel Vectors

Naoya Nomura[1]($\boxtimes$), Akihiro Fujii[1], Teruo Tanaka[1], Kengo Nakajima[2], and Osni Marques[3]

[1] Kogakuin University, Tokyo, Japan
em15016@ns.kogakuin.ac.jp
[2] The University of Tokyo, Tokyo, Japan
[3] Lawrence Berkeley National Laboratory, Berkeley, USA

Abstract. The smoothed aggregation algebraic multigrid (SA-AMG) method is among the fastest solvers for large-scale linear equations. It achieves good convergence by generating small matrices from the original matrix problem. However, the convergence of the method can be further improved by using near-kernel vectors. Our research investigates the effectiveness of using multiple near-kernel vectors and finds the near-kernel vectors that are most important for obtaining rapid convergence. We apply our method to the three-dimensional problem in elasticity. The known near-kernel vectors (the parallel translation and rotation vectors) improve the convergence and execution time of the SA-AMG method. We use an iterative process known as the V-cycle to extract multiple near-kernel vectors. In numerical experiments, we show that a suitable choice of the near-kernel vectors reduces the number of iterations by up to two-thirds and halves the execution time, compared to use of the known near-kernel vectors. Our method will be effective for cases in which the same matrix problem is solved repeatedly.

Keywords: Linear solver · Algebraic multigrid method · Near-kernel vectors · Performance evaluation

1 Introduction

Iterative solutions to large-scale systems of linear equations $Ax = b$ are often required in scientific computing. Among the fastest solvers for these equations is the algebraic multigrid (AMG) method [1], a multilevel method that creates smaller matrices from the matrix problem. A variant called the smoothed aggregation AMG (SA-AMG) method [2–4] is effective for solving various problems and is widely used.

The SA-AMG method comprises a setup part and a solution part. The setup part creates a graph structure based on a matrix problem and defines a coarse problem based on aggregates of unknowns. Recursive application of this process generates multiple small matrices. The solution part repeatedly applies relaxation (e.g., the Jacobi method) to the hierarchical matrices constructed in the

© Springer International Publishing AG 2017
I. Dutra et al. (Eds.): VECPAR 2016, LNCS 10150, pp. 52–63, 2017.
DOI: 10.1007/978-3-319-61982-8_7

setup part, thereby achieving rapid convergence. The structure is hierarchical, with fine levels (large matrices) and coarser levels (small matrices). The finest level corresponds to the original matrix problem.

The SA-AMG method can incorporate error components that are difficult to correct using ordinary relaxation methods. These error components typically correspond to the near-kernel vectors, defined as nonzero vectors x satisfying $Ax \approx 0$. The SA-AMG method sets these error components and efficiently corrects them by moving them to coarser levels.

In this research, we focus on the three-dimensional problem in elasticity, in which the near-kernel vectors are the parallel translation and rotation vectors. The use of these near-kernel vectors improves the convergence of the SA-AMG method in the target problem. To our knowledge, methods for extracting near-kernel vectors and their efficiency have been rarely reported other than for αSA [5], which is described in Sect. 4. This study reports a simple method for using V-cycles to extract multiple near-kernel vectors and numerically evaluates its performance.

2 SA-AMG Method

This section describes the SA-AMG method, which creates and solves hierarchical matrices.

The AMG method is among the fastest solvers for large-scale linear equations. It creates multiple small matrices from the matrix problem and uses their solutions to solve the problem settings. The AMG method has setup part and a solution part. Figure 1 shows the setup part, which creates coarser matrix problems and the *prolongation* and *restriction* matrices from the fine matrix problem, and it does this hierarchically. The top level is the original matrix problem, and progressively coarser levels are represented by progressively smaller matrices. The number of levels depends on the size of the matrix problem. Data are moved between levels by the interlevel prolongation and restriction matrices.

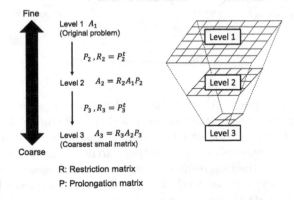

Fine

Level 1 A_1
(Original problem)

$P_2, R_2 = P_2^t$

Level 2 $A_2 = R_2 A_1 P_2$

$P_3, R_3 = P_3^t$

Level 3 $A_3 = R_3 A_2 P_3$
(Coarsest small matrix)

Coarse

Level 1

Level 2

Level 3

R: Restriction matrix

P: Prolongation matrix

Fig. 1. Setup part of the SA-AMG

The coarse matrices and the interlevel prolongation and restriction matrices are used in the solution. Figure 2 shows the solution part of the AMG method, which performs matrix vector multiplication and relaxation. First, the relaxation method is applied to the finest matrix. Next, the residual vector is calculated and multiplied by the restriction matrix, which moves it to a coarser level. At the coarser level, this vector is set as the right-hand vector. After applying relaxation at the coarser level, the corrected vector is determined and multiplied by the prolongation matrix, which moves it to a finer level. The solution is added to the existing solution vector at the finer level. Finally, relaxation is repeated at the finer level. This process is called the V-cycle.

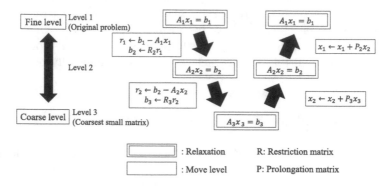

Fig. 2. V-cycle of the SA-AMG (Solution part)

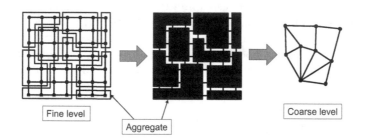

Fig. 3. How an aggregate is made

There are many types of AMG methods with different ways of creating the interlevel matrices [6]. In this study, we used the SA-AMG method, which creates coarser matrices from a graph structure that is based on the matrix problem. The unknowns and nonzero elements correspond to nodes and edges, respectively. The SA-AMG method then aggregates the unknowns and assigns each aggregate to a node at the next coarser level. Each node at a finer level belongs to a corresponding aggregate. Figure 3 shows the making of an aggregate. The SA-AMG method creates aggregates using the graph structure based on the matrix problem, and each aggregate of unknowns at a finer level corresponds to the unknowns at a coarser level. For interpolation, the SA-AMG method sets weight values on each of the aggregate nodes. The weights are stored in an interlevel matrix. The coarser matrix is calculated by two matrix-matrix multiplications RAP, where R and P are the restriction and prolongation matrices, respectively. To improve the convergence of the SA-AMG method, the interlevel matrix can be constructed from the near-kernel vectors.

3   Near-Kernel Vector

In this section, we define near-kernel vectors and explain how they are determined in the SA-AMG method.

The near-kernel vector is a vector x that satisfies $Ax \approx 0 (x \neq 0)$. When the equation $Ax = b$ is solved by a regular iterative solver, the solution vector is updated using the vector b. However, the error components of the near-kernel vectors cannot be corrected by this vector. Consequently, the convergence stagnates.

For moving the near-kernel vector components to coarser levels, the SA-AMG method uses the near-kernel vectors to calculate the interlevel operators. Thus, these components are efficiently corrected at coarser levels, and rapid convergence is achieved [2–4]. In some cases, the near-kernel vectors can be determined from the problem settings. For example, the near-kernel vectors for a problem in elasticity are the translation and rotation vectors.

Figure 4 illustrates the use of near-kernel vectors in the interlevel matrix. In this figure, the prolongation matrix is constructed from two near-kernel vectors and two aggregates. In particular, the prolongation matrix is created by selecting the corresponding elements from the near-kernel vectors for each aggregate. In the SA-AMG method, the number of columns in the prolongation matrix is proportional with the number of near-kernel vectors, and this increases the calculation costs. Therefore, there is a trade-off between the number of near-kernel vectors and the execution time.

Figures 5 and 6 show how the number of iterations necessary for convergence depends on the number of near-kernel vectors. Figure 5 shows the target problem of this experiment. The matrix structure of this problem in elasticity is described

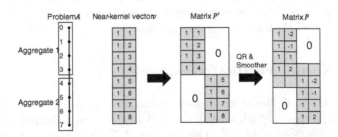

Fig. 4. Construction of the prolongation matrix from the near-kernel vectors

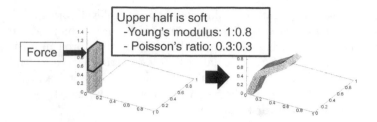

Fig. 5. Experiment with multiple near-kernel vectors

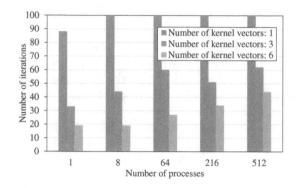

Fig. 6. Effect of multiple near-kernel vectors on the required number of iterations in the SA-AMG method

in Sect. 5.1; the object comprises a soft upper part and a hard lower part. A force is applied over a small area on the upper side. Young's modulus is 1:0.8, and Poisson's ratio is 0.3:0.3. The graph in Fig. 6 shows the number of iterations necessary to reach convergence. The number of near-kernel vectors is indicated; "Number of kernel vectors: 1" uses a constant vector, "Number of kernel vectors: 3" uses only translation vectors, and "Number of kernel vectors: 6" uses both translation and rotation vectors. This experiment is a weak scaling test (with a local domain size per process of $6 \times 15 \times 60$). The domain size is proportional to the number of processes. We note that using more near-kernel vectors reduces the number of iterations because the near-kernel vectors are already known from the problem settings. In the next section, we consider the extraction of additional near-kernel vectors that are not known a priori.

4   Near-Kernel Vector Extraction

This section describes the method used to extract near-kernel vectors.

The αSA method, which uses V-cycles to extract near-kernel vectors, was proposed in a previous study [5]. The authors of [5] first calculated the coarser-level near-kernel vectors and then interpolated them to the finest level, for use as near-kernel vectors at that level. However, we perform multiple V-cycle iterations in order to directly calculate the near-kernel vectors, as follows:

1. Initialize the vector x randomly.
2. Iterate the V-cycle μ times to solve $Ax = 0$.
3. Set the solution vector x in Step 2 as an additional near-kernel vector. If the number of near-kernel vectors is insufficient, return to Step 1.
4. Output the extracted near-kernel vectors.

After solving $Ax = 0$ by performing a sufficient number of V-cycle iterations, we are left with the near-kernel error component, which cannot be solved by a V-cycle. The V-cycle iterations are executed on near-kernel vectors extracted in the previous iteration steps. Thus, the new near-kernel vector is different from the previous near-kernel vectors. We note that in order to perform a V-cycle, it is necessary to have an initial near-kernel vector, and that vector must be determined from the problem settings. In subsequent processes, the method extracts independent near-kernel vectors.

In Step 2, the parameter μ is provided as an input. This parameter specifies the number of V-cycle iterations and largely determines the performance of the SA-AMG method. As an area of future work, we will consider the optimization of μ. In the present study, we set μ to 20.

5   Numerical Experiments

5.1   Experimental Environments and Problem Setup

Experiments were performed on an FX10 supercomputer system (Oakleaf-FX) [7] at the University of Tokyo. Each node of the FX10 is equipped with one SPARC64 IXfx processor (1.848 GHz, 16 cores) and 32 GB memory. These nodes are connected through the 6D mesh/tours network (5 GB/s/link, bidirectional). We launched one process per core (Flat MPI model), employing up to 512 cores.

The experimental subject was a three-dimensional problem in elasticity, in which an elastic object is pressed under a constant force, and the displacements of all parts of the object are to be determined. The problem was constructed as a 3×3 block matrix at each node. This linear elasticity problem in a simple cubic geometry in a heterogeneous media was solved using the finite-element method (FEM). Tri-linear hexahedral (cubic) elements were used for the discretization, and a heterogeneous distribution of Young's modulus in each element was calculated by a sequential Gauss algorithm, which is widely used in the area of geostatistics [8]. The object consisted of an inner hard cube and an outer soft cube, and a force was applied to a small area on the upper face (see Fig. 7). Young's modulus (indicating the stiffness of the material) was 5 in the hard area and 0.5 in the soft area, and Poisson's ratio was 0.3 in all areas. Moreover, the experiment was performed as a weak scaling (with a local domain per process of $15 \times 15 \times 15$). The problem domain was divided into several subdomains with equal intervals on each axis.

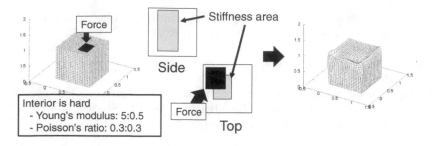

Fig. 7. Setup of the experimental problem in elasticity

The experiment was implemented in the AMGS library [9], which solves large-scale linear equations by the AMG method. The solution part was executed by the generalized product bi-conjugate gradient (GPBiCG) method [10], and one iteration of the V-cycle was used as a preconditioner. We implemented the program contained in the AMGS library, and we implemented the GPBiCG method in the Fortran programming language with the MPI library. We used the Fujitsu Fortran compiler with the option "-Kfast,openmp". We linked to the MPI library provided by Fujitsu.

The relaxation procedure of the solution part was performed twice per level by the Symmetric Gauss-Seidel method, ignoring the data dependency beyond the border of the processing domain. The coarsest level was determined to contain less than 100 unknown blocks; that is, there were fewer than $100 \times k$ unknowns at the coarsest level when it is initialized with k near-kernel vectors. The termination criterion for the 2-norm of the relative residuals was set to 1.0×10^{-7}. The maximum number of iterations was set to 500.

5.2   Experimental Results

In this experiment, we investigated the performance of the SA-AMG method by changing the number of near-kernel vectors. The near-kernel vector setups are provided in Table 1. The near-kernel vectors in the extraction process were initialized as the parallel translation vectors.

Figure 8 shows the results. The five graphs correspond to results with 1, 8, 64, 216, and 512 processes, respectively. The horizontal axis indicates the number of near-kernel vectors that were used. The height of each bar indicates the time

Table 1. Experimental setup

Near-kernel vectors	Details
3 provided	Parallel translation in each axis direction (X,Y,Z)
6 provided	Parallel translation + rotation on each axis (X,Y,Z)
3p+1, 3p+2,...	Parallel translation + extracted near-kernel vectors (up to 7)

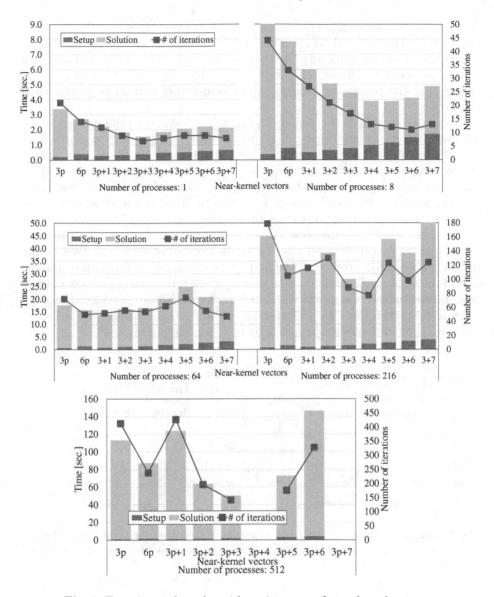

Fig. 8. Experimental results with various sets of near-kernel vectors

until convergence, while the line indicates the required number of iterations. "Setup" and "Solve" indicate the execution times of the setup part and solution part, respectively. If the number of iterations exceeded 500, the execution time is not shown. In this experiment, we disregarded the time required to extract the near-kernel vectors. As shown in Fig. 8, the lowest number of iterations and the best execution time were obtained when 3 or 6 near-kernel vectors were provided. However, using additional near-kernel vectors does not always improve the rate of convergence. For example, when running 512 processes, the method with 3p+7 near-kernel vectors failed to converge. In 64 processes, the lowest number of iterations was achieved for 3p+7 near-kernel vectors, whereas 3p+1 near-kernel vectors achieved the lowest execution time. This result can be explained by noting that the processing time for 3p+7 near-kernel vectors is greater than that for 3p+1.

Figure 9 shows the best results from Fig. 8. The left and right panels show the number of iterations and the execution time, respectively. The horizontal axis indicates the number of processes, and the vertical axis indicates the number of iterations (left panel) and the execution time (right panel). "3 provided" and "6 provided" indicate the results obtained when using "parallel translation" and "parallel translation + rotation" vectors, respectively. "Best in extracted vectors" is the result that had the best execution time of the 7 sets of extracted vectors that were considered. This figure shows that by appropriately determining the near-kernel vectors, we can dramatically reduce the number of iterations. Even in the largest problem (512 processes), the best of the extracted vectors approximately halved the number of iterations and reduced the execution time by approximately 40%, relative to the case in which 6 near-kernel vectors were provided. However, we note that when there are 512 processes, many more iterations (iterations = 141) are required than when there are 64 processes (iterations = 47).

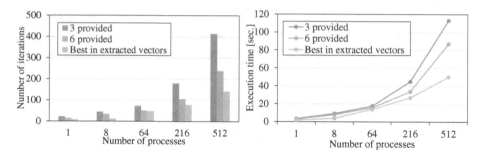

Fig. 9. Number of iterations (left) and execution time (right) for various sets of near-kernel vectors (3 provided, 6 provided, and the best number of extracted near-kernel vectors)

In order to further evaluate this approach, a simple problem was prepared. For both parts (inner and outer) shown in Fig. 7, Young's modulus was changed to unity (that is, it was made uniform throughout the system). Figure 10 shows the number of iterations required for convergence; note that it was nearly the same for both 64 (iterations = 11) and 512 (iterations = 13) processes. Tuning the number of extracted near-kernel vectors reduces the number of iterations even for this easy problem.

These results have shown that the number of iterations can be reduced by the choice of a suitable number of near-kernel vectors to be extracted. For the

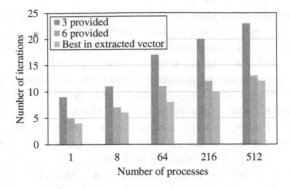

Fig. 10. Number of iterations for various near-kernel vectors in an easy problem (Young's modulus is changed to 1:1)

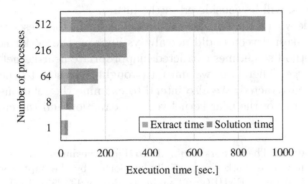

Fig. 11. Search time for best near-kernel vectors

problem described in Fig. 7, we now consider the cost of extracting near-kernel vectors and the optimal number to use. Figure 11 shows the time required to extract near-kernel vectors and to find the solution, for each of the number of vectors from $3p+1$ to $3p+7$. The horizontal and vertical axes correspond to the execution time and the number of processes, respectively. "Extract time" is the time required to extract 7 vectors, and "Solution time" is the time required to solve the problem 7 times, each with a different set of near-kernel vectors. For the example with 216 processes, as shown in Fig. 11, the time to extract the near-kernel vectors and to search is only about 9 times that with 6p near-kernel vectors. Thus, this method is an efficient way to repeatedly solve the same matrix equation with different initial conditions.

6   Conclusion

In this paper, we investigated the effect on the performance of the SA-AMG method of using multiple V-cycle iterations to extract multiple near-kernel vectors. To do this, we applied our method to a three-dimensional problem in elasticity. The results were compared with those when using the known vectors (the parallel translation and rotation vectors). In the largest problem (512 processes), using the extracted near-kernel vectors halved the number of iterations and decreased the solution time by approximately 40%, relative to the results when using the ordinary translation and rotation vectors. Thus, the ordinary V-cycle iterations were found to extract near-kernel vectors that were effective for rapid convergence. Another problem is the scalability of this for cases with many nodes. Figure 9 shows that both the number of iterations and the computation time increase as the problem size increases. This occurs primarily because of the localized block-Jacobi-type Gauss-Seidel smoothers in the SA-AMG procedure. The increase in the number of iterations is very significant for ill-conditioned problems, but it is not very large when the condition number is close to unity, as shown in Fig. 6. Stabilization of the localized smoother is a very critical issue. Some remedies described in Ref. [11], such as extending the overlapped zones, will be considered in the future.

We conclude with some remarks on directions for further research. First, setting many near-kernel vectors did not always improve the performance; in particular, the method sometimes extracted inappropriate near-kernel vectors that failed to converge. Therefore, we must thoroughly investigate the near-kernel vectors that are extracted. We also intend to examine the relationship between the residual history of the near-kernel vector extraction and the effectiveness of these vectors.

Acknowledgments. The authors would like to thank the anonymous referees for their valuable comments. This work was partially supported by the Japan Society for the Promotion of Science KAKENHI (grant numbers 15K15998, 25330144), and Initiative on Promotion of Supercomputing for Young or Women Researchers, Supercomputing Division, Information Technology Center, The University of Tokyo.

References

1. Pereira, F.H., Verardi, S.L.L., Nabeta, S.I.: A fast algebraic multigrid preconditioned conjugate gradient solver. Appl. Math. Comput. **179**, 344–351 (2006)
2. Vanek, P., Brezina, M., Mandel, J.: Convergence of algebraic multigrid based on smoothed aggregation. Numer. Math. **88**, 559–579 (2001)
3. Vanek, P., Mandel, J., Brezina, M.: Algebraic multigrid by smoothed aggregation for second and fourth order elliptic problems. Computing **56**, 179–196 (1998)
4. Chan, T.F., Vanek, P.: Multilevel algebraic Elliptic Solvers, UCLA Math, Dept., CAM Report (1999)
5. Brezina, M., Falgout, R., Maclachlan, S., Manteuffel, T., Mccormick, S., Ruge, J.: Adaptive smoothed aggregation (αSA). SIAM J. Sci. Comput. **25**(6), 1896–1920 (2004)
6. Fujii, A., Oyanagi, Y.: Evaluation of algebraic multi-grid method: an efficient linear solver for scientific simulations. Simulations **28**(4), 149–154 (2009). 2009–12-15, pp. 9–14
7. Information Technology Center: The University of Tokyo. http://www.cc.u-tokyo.ac.jp/

8. Deutsch, C.V., Journel, A.G.: GSLIB Geostatistical Software Library and User's Guide, 2nd edn. Oxford University Press, Oxford (1998)
9. AMGS Library: http://hpcl.info.kogakuin.ac.jp/lab/software/amgs
10. Zhang, S.-L.: GPBi-CG: generalized product-type methods based on Bi-CG for solving nonsymmetric linear systems SIAM. J. Sci. Comput. **18**(2), 537–551 (1997)
11. Nakajima, K.: Strategies for preconditioning methods of parallel iterative solvers in finite-element applications on geophysics. Advances in Geocomputing. Lecture Notes in Earth Science, vol. 119, pp. 65–118. Springer, Heidelberg (2009)

Computing the Bidiagonal SVD Through an Associated Tridiagonal Eigenproblem

Osni Marques[1(✉)] and Paulo B. Vasconcelos[2(✉)]

[1] Lawrence Berkeley National Laboratory, Berkeley, USA
oamarques@lbl.gov
[2] Faculdade de Economia and CMUP, Universidade Do Porto, Porto, Portugal
pjv@fep.up.pt

Abstract. In this paper, we present an algorithm for the singular value decomposition (SVD) of a bidiagonal matrix by means of the eigenpairs of an associated symmetric tridiagonal matrix. The algorithm is particularly suited for the computation of a subset of singular values and corresponding vectors. We focus on a sequential implementation of the algorithm, discuss special cases and other issues. We use a large set of bidiagonal matrices to assess the accuracy of the implementation and to identify potential shortcomings. We show that the algorithm can be up to three orders of magnitude faster than existing algorithms, which are limited to the computation of a full SVD.

1 Introduction

It is well known that the singular value decomposition (SVD) of a matrix $A \in \mathbb{R}^{m \times n}$, namely $A = USV^T$, with left singular vectors $U = [u_1, u_2, \ldots u_n]$, right singular vectors $V = [v_1, v_2, \ldots v_n]$, and singular values $S = diag(s_1, s_2, \ldots s_n)$, $s_1 \geq s_2 \geq \ldots s_n \geq 0$, can be obtained through the eigenpairs (λ, x) of the matrices $C_{n \times n} = A^T A$ and $C_{m \times m} = AA^T$. However, if A is square and orthogonal $C_{n \times n}$ and $C_{m \times m}$ are both the identity and provide little information about the singular vectors of A, which are not unique: $A = (AV)IV^T$ is the SVD of A for any orthogonal matrix V. A potential difficulty for some algorithms (e.g. the one presented in this paper) is the existence of large clusters of close singular values, as this may have an impact on the orthogonality of the computed singular vectors.

Alternatively to $C_{n \times n}$ and $C_{m \times m}$, the SVD can be obtained through the augmented matrix [1]

$$C = \begin{bmatrix} 0 & A \\ A^T & 0 \end{bmatrix} = J \begin{bmatrix} -S & 0 \\ 0 & S \end{bmatrix} J^T, \quad J = \begin{bmatrix} U & U \\ -V & V \end{bmatrix} / \sqrt{2}, \tag{1}$$

such that the eigenvalues of C are $\pm s$ and its eigenvectors are mapped into the singular vectors of A (scaled by $\sqrt{2}$) in a very structured manner.

In practical calculations, the SVD of a full matrix A involves the reduction of A to bidiagonal form B through orthogonal transformations, i.e. $A = \hat{U} B \hat{V}^T$.

© Springer International Publishing AG 2017
I. Dutra et al. (Eds.): VECPAR 2016, LNCS 10150, pp. 64–74, 2017.
DOI: 10.1007/978-3-319-61982-8_8

The singular values are thus preserved; the singular vectors of B need to be back transformed into those of A.

If B is an upper bidiagonal matrix with $(a_1, a_2, \ldots a_n)$ on the main diagonal and $(b_1, b_2, \ldots b_{n-1})$ on the off diagonal, we can replace A with B in (1) to obtain $C = P\, T_{GK}\, P^T$, where T_{GK} is the Golub-Kahan symmetric tridiagonal matrix,

$$T_{GK} = tridiag \begin{pmatrix} a_1 & b_1 & a_2 & b_2 \ldots b_{n-1} & a_n \\ 0 & 0 & 0 & 0 \quad \ldots & 0 \quad 0 \\ a_1 & b_1 & a_2 & b_2 \ldots b_{n-1} & a_n \end{pmatrix}, \tag{2}$$

and the perfect shuffle $P = [e_{n+1}, e_1, e_{n+2}, e_2, e_{n+3}, \ldots e_{2n}]$, where the e's are the columns of the identity matrix of dimension $2n$. Then, if the eigenpairs of T_{GK} are $(\pm s, z)$, with $\|z\| = 1$, and from (1), we obtain $z = P(u^T, \pm v^T)/\sqrt{2}$ [6]. Thus, we can extract the SVD of B from the eigendecomposition of T_{GK}.

Table 1 lists the current LAPACK subroutines intended for the computation of the SVD of bidiagonal matrices, and eigenvalues and eigenvectors of tridiagonal matrices. The trade-offs (performance, accuracy) of these eigensolvers have been thoroughly examined in [3]. We are interested in how the symmetric tridiagonal (ST) subroutines could be applied to (2), specially for the computation of subsets of eigenpairs, which in turn could reduce the computational costs when a full SVD is not needed (or for the computations of subsets in parallel). While STEDC could be potentially redesigned to compute a subset of eigenvectors, saving some work but only at the top level of recursion of the divide-and-conquer algorithm, STEVX and STEMR offer more straightforward alternatives. STEVX performs bisection to find selected eigenvalues followed by inverse iteration to find their eigenvectors, for an $O(n)$ cost per eigenpair. STEVX can occasionally fail to provide orthogonal eigenvectors when the eigenvalues are too closely clustered. In contrast, STEMR uses a much more sophisticated algorithm called MRRR [4,5] to guarantee orthogonality. An improved version of the MRRR algorithm targeting T_{GK} in order to compute the SVD has been proposed in [6]; however, our experiments with an implementation given in [6] produced vectors with inadequate level of orthogonality, for relatively simple matrices. (We illustrate with one case in the numerical experiments session.) Therefore, we have decided to adopt STEVX for computing eigenvalues and eigenvectors of (2), even though it has known failure modes that we discuss later.

Table 1. LAPACK's (bidiagonal, BD) SVD and (tridiagonal, ST) eigensolvers.

Routine	Usage	Algorithm
BDSQR	all s and (opt.) u and/or v	implicit QL or QR
BDSDC	all s and (opt.) u and v	divide-and-conquer
STEQR	all λ's and (opt.) x	implicit QL or QR
STEVX	selected λ's and (opt.) x	bisection & inverse iteration
STEDC	all λ's and (opt.) x	divide-and-conquer
STEMR	selected λ's and (opt.) x	MRRR

The main contribution of this paper is to discuss an implementation of an algorithm for the SVD of a bidiagonal matrix obtained from eigenpairs of a tridiagonal matrix T_{GK}. This implementation is called BDSVDX, introduced in LAPACK 3.6.0. While the associated formulation is not necessarily new, as mentioned above, its actual implementation requires care in order to deal correctly with multiple or tightly clustered singular values, or some cases of splitting. To the best of our knowledge, no such implementation has been done and exhaustively tested. In concert with BDSVDX we have also developed GESVDX, which takes a general matrix A, reduces it to bidiagonal form B, invokes BDSVDX, and then maps the output of BDSVDX into the SVD of A. In LAPACK, the current counterparts of GESVDX are GESVD and GESDD, which are based on the BD subroutines listed in Table 1 and can only compute all singular values (and optionally singular vectors). This can be much more expensive if only a few singular values and vectors are desired.

The rest of the paper is organized as follows. First, we discuss how singular values are mapped into the eigenvalue spectrum. Then, we discuss special cases, the criterion for splitting a bidiagonal matrix, and other implementation details. Next, we show the results of our tests with BDSVDX using a large set of bidiagonal matrices, to assess both accuracy and computational performance. We compare the performances of BDSQR, BDSDC and BDSVDX, and GESVD, GESDD and GESVDX. Finally, we discuss limitations and opportunities for future work.

2   Mapping Singular Values into Eigenvalues

Similarly to BDSQR and BDSDC, BDSVDX allows the computation of singular values only or singular values and the corresponding singular vectors. Borrowing features from STEVX, BDSVDX can be used in three modes, through a character variable RANGE. If RANGE = "A", all singular values will be found: BDSVDX will compute the smallest (negative or zero) n eigenvalues of the corresponding T_{GK}. If RANGE = "V", all singular values in the half-open interval (VL,VU] will be found: BDSVDX will compute the eigenvalues of the corresponding T_{GK} in the interval $(-VU, -VL]$. If RANGE = "I", the IL-th through IU-th singular values will be found: the indices IL and IU are mapped into values (similar to VL and VU) by applying bisection to T_{GK}. VL, VU, IL and IU are arguments of BDSVDX (which are mapped into similar arguments for STEVX).

For a bidiagonal matrix B of dimension n, if singular vectors are requested, BDSVDX returns an array Z of dimension $2n \times p$, where $p \leq n$ is a function of RANGE. Each column of Z will contain $(u_i^T, v_i^T)^T$ corresponding to singular value s_i, i.e. (using MATLAB notation) $Z = [\, U;\ V\,]$. STEVX returns eigenvalues (and corresponding vectors) in ascending order, so we target the negative part of the eigenvalue spectrum (i.e. $-S$) in (1). Therefore, the absolute values of the returned eigenvalues give us the singular values in the desired order, $s_1 \geq s_2 \geq \ldots s_n \geq 0$. We only need to change the signs of the entries in the eigenvectors that are reloaded to V. We note that BDSVDX inherits some shortcomings from STEVX: in extreme situations bisection may fail to converge, or not all eigenvalues with indices IL:IU can be found, or inverse iteration fails to converge after the allowed number of iterations is reached. However, we have never observed the occurrence of any of these anomalies in our thorough tests.

3 Splitting: Special Cases

The criterion for splitting in BDSVDX is the same that is used in STEQR and is discussed in [7]. We first form the matrix T_{GK} and check for splitting in two phases, starting with the off diagonal entries of T_{GK} with even indices (i.e. the b's). If, for a given i, $b_i = 0$ (or it is tiny enough to be set to zero) the matrix B splits and the SVD for each resulting (square) submatrix of B can be obtained independently. This effect is propagated into the associated T_{GK}, i.e. the eigenvalues and eigenvectors of each submatrix of T_{GK} can be obtained independently. We then check the off diagonal entries of T_{GK} with odd indices (i.e. the a's). If, for a given j, $a_j = 0$ (or tiny enough), we end up with rectangular bidiagonal matrices, which do not have equal numbers of left and right singular vectors. This complicates our simple approach for extracting singular vectors of B from eigenvectors of T_{GK}. The problem can be reduced to one of the three special cases illustrated below with small matrices.

Zero in the interior. If $n = 5$ and $a_3 = 0$, we have the following SVD:

$$B = bidiag \begin{pmatrix} & b_1 & b_2 & b_3 & b_4 \\ a_1 & a_2 & 0 & a_4 & a_5 \end{pmatrix} = \begin{bmatrix} U_1 & \\ & U_2 \end{bmatrix} \begin{bmatrix} S_1 & \\ & S_2 \end{bmatrix} \begin{bmatrix} V_1^T \\ V_2^T \end{bmatrix},$$

where U_1 and V_2 are 2-by-2, U_2 and V_1 are 3-by-3, S_1 is 2-by-3 (its third column contains only zeros), and S_2 is 3-by-2 (its third row contains only zeros). If we construct $T_{GK}^{(1)}$ and $T_{GK}^{(2)}$ matrices as

$$T_{GK}^{(1)} = tridiag \begin{pmatrix} & a_1 & b_1 & a_2 & b_2 \\ 0 & 0 & 0 & 0 & 0 \\ & a_1 & b_1 & a_2 & b_2 \end{pmatrix}, \quad T_{GK}^{(2)} = tridiag \begin{pmatrix} & b_3 & a_4 & b_4 & a_5 \\ 0 & 0 & 0 & 0 & 0 \\ & b_3 & a_4 & b_4 & a_5 \end{pmatrix},$$

then the first three columns of their respective eigenvector matrices are

$$Z_{5\times3}^{(1)} = \begin{bmatrix} v_{1,1}^{(1)} & v_{1,2}^{(1)} & v_{1,3}^{(1)} \\ u_{1,1}^{(1)} & u_{1,2}^{(1)} & 0 \\ v_{2,1}^{(1)} & v_{2,2}^{(1)} & v_{2,3}^{(1)} \\ u_{2,1}^{(1)} & u_{2,2}^{(1)} & 0 \\ v_{3,1}^{(1)} & v_{3,2}^{(1)} & v_{3,3}^{(1)} \end{bmatrix} D^{-1}, \quad Z_{5\times3}^{(2)} = \begin{bmatrix} u_{1,1}^{(2)} & u_{1,2}^{(2)} & u_{1,3}^{(2)} \\ v_{1,1}^{(2)} & v_{1,2}^{(2)} & 0 \\ u_{2,1}^{(2)} & u_{2,2}^{(2)} & u_{2,3}^{(2)} \\ v_{2,1}^{(2)} & v_{2,2}^{(2)} & 0 \\ u_{3,1}^{(2)} & u_{3,2}^{(2)} & u_{3,3}^{(2)} \end{bmatrix} D^{-1}$$

where $Z_{5\times3}^{(1)}$ and $Z_{5\times3}^{(2)}$ show how the entries of the eigenvectors corresponding to the three smallest (negative) eigenvalues of $T_{GK}^{(1)}$, $\lambda_1^{(1)} < \lambda_2^{(1)} < \lambda_3^{(1)}$, and $T_{GK}^{(2)}$, $\lambda_1^{(2)} < \lambda_2^{(2)} < \lambda_3^{(2)}$ relate to the entries of U_1, U_2, V_1 and V_2, where $v_{ij}^{(1)}$ are the entries of V_1 and so on. Note that the left and right singular vectors corresponding to s_3 are in different matrices, with $D = diag(\sqrt{2}, \sqrt{2}, 1)$. (The array Z returned by BDSVDX would be, in MATLAB notation, $Z = [\ Z^{(1a)}_{5\times2}\ \ Z^{(1b)}_{5\times3};\ 0\ \ Z^{(2)}_{5\times3}\ ]$, where $Z^{(1a)}_{5\times2}$ contains the first two columns of $Z^{(1)}_{5\times3}$, while $Z^{(1b)}_{5\times3}$ has zeros in its two first columns and the last column of $Z^{(1)}_{5\times3}$ in its last column.)

Zero at the top. If $n = 4$ and $a_1 = 0$, we have the following SVD:

$$B = bidiag \begin{pmatrix} & b_1 & b_2 & b_3 \\ 0 & a_2 & a_3 & a_4 \end{pmatrix} = [U] \begin{bmatrix} 0 \\ & S \end{bmatrix} \begin{bmatrix} 1 \\ & V^T \end{bmatrix},$$

where U is 4-by-4, S is 3-by-3, and V is 3-by-3. If we construct a T_{GK} from B, its first row and column will be zero, and the entries of the eigenvectors corresponding to the five smallest eigenvalues of T_{GK} (again, related explicitly to singular values of B) relate to the entries of U and V as shown in Table 2. (The array Z returned by BDSVDX would be formed by taking the last four columns of $Z^{(1)}_{8\times5}$; its last column is concatenated with the first column of $Z^{(1)}_{8\times5}$.)

Zero at the bottom. If $n = 4$ and $a_4 = 0$, we have the following SVD:

$$B = bidiag \begin{pmatrix} b_1 & b_2 & b_3 \\ a_1 & a_2 & a_3 & 0 \end{pmatrix} = \begin{bmatrix} U \\ & 1 \end{bmatrix} \begin{bmatrix} S \\ & 0 \end{bmatrix} [V^T],$$

where U is 3-by-3, S is 3-by-3, and V is 4-by-4. If we construct a T_{GK} from B, its last row and column will be zero, the entries of the eigenvectors corresponding to the five smallest eigenvalues of T_{GK} (again, related explicitly to singular values of B) relate to the entries of U and V as shown in Table 2. (The array Z returned by BDSVDX would be formed by taking the first four columns of $Z^{(2)}_{8\times5}$; its last column is concatenated with the last column of $Z^{(2)}_{8\times5}$.)

If the eigenpairs of T_{GK} are $(\pm s, z)$, with $\|z\| = 1$, and from (1), we obtain $z = P(u^T, \pm v^T)/\sqrt{2}$ [6]. Thus, we can extract the SVD of B from the eigendecomposition of T_{GK}.

Table 2. Relation between the eigenvectors of T_{GK} and the entries of U and V for a zero at the top or bottom of B.

Zero at the top	Zero at the bottom
$Z^{(1)}_{8\times5} = \begin{bmatrix} 1 \\ & Z^{(1)}_{7\times4} \end{bmatrix} (D^{(1)})^{-1}$	$Z^{(2)}_{8\times5} = \begin{bmatrix} Z^{(2)}_{7\times4} \\ & 1 \end{bmatrix} (D^{(2)})^{-1}$
$D^{(1)} = diag(1, \sqrt{2}, \sqrt{2}, \sqrt{2}, 1)$	$D^{(2)} = diag(\sqrt{2}, \sqrt{2}, \sqrt{2}, 1, 1)$
Columns of $Z^{(1)}_{7\times4}$:	Columns of $Z^{(2)}_{7\times4}$:
$(u_{1,1}\ \ v_{1,1}\ \ u_{2,1}\ \ v_{2,1}\ \ u_{3,1}\ \ v_{3,1}\ \ u_{4,1})^T$	$(v_{1,1}\ \ u_{1,1}\ \ v_{2,1}\ \ u_{2,1}\ \ v_{3,1}\ \ u_{3,1}\ \ v_{4,1})^T$
$(u_{1,2}\ \ v_{1,2}\ \ u_{2,2}\ \ v_{2,2}\ \ u_{3,2}\ \ v_{3,2}\ \ u_{4,2})^T$	$(v_{1,2}\ \ u_{1,2}\ \ v_{2,2}\ \ u_{2,2}\ \ v_{3,2}\ \ u_{3,2}\ \ v_{4,2})^T$
$(u_{1,3}\ \ v_{1,3}\ \ u_{2,3}\ \ v_{2,3}\ \ u_{3,3}\ \ v_{3,3}\ \ u_{4,3})^T$	$(v_{1,3}\ \ u_{1,3}\ \ v_{2,3}\ \ u_{2,3}\ \ v_{3,3}\ \ u_{3,3}\ \ v_{4,3})^T$
$(u_{1,4}\ \ 0\ \ u_{2,4}\ \ 0\ \ u_{3,4}\ \ 0\ \ u_{4,4})^T$	$(v_{1,4}\ \ 0\ \ v_{2,4}\ \ 0\ \ v_{3,4}\ \ 0\ \ v_{4,4})^T$

4   Reorthogonalization of Vectors

As discussed earlier, given an eigenvector z_i of T_{GK}, $z_i = P(u_i^T, -v_i^T)^T/\sqrt{2}$ ($i \leq 1 \leq n$). We could simply create $\hat{u}_i$ with the even entries of z_i and $\hat{v}_i$ with the odd entries of z_i and multiply those vectors by $\sqrt{2}$ in order to obtain u_i and v_i. However, in our implementation we explicitly normalize $\hat{u}_i$ and $\hat{v}_i$. This allows us to check how far the norms of $\hat{u}_i$ and $\hat{v}_i$ are from $\frac{1}{\sqrt{2}}$, which may be the case if z_i is associated with a small λ. Then, if needed, we apply a Gram-Schmidt reorthogonalization to $\hat{u}_i$ and $\hat{v}_i$. Our test for triggering a reorthogonalization is based on $|\|\hat{u}\| - \frac{1}{\sqrt{2}}| \geq tol$ (similarly for $\hat{v}$), $tol = \sqrt{\varepsilon}$, where ε is the machine precision. However, we have identified matrices for which this test is not sufficient, which suggests the need for a strategy that takes into account the separation of λ's. This is the case, for example, of the bidiagonal matrix defined as $a_i = 10^{-(2i-1)}, i = 1, 2, \ldots n$, $b_i = 10a_i, i = 1, 2, \ldots n - 1, n = 8$, for which $s_1 \approx 1.005$, $s_7 \approx 10^{-12}$ and $s_8 \approx 10^{-22}$. While the eigenvectors of T_{GK} as computed by STEVX are orthogonal to work precision, specifically, $\mathcal{Z} = [z_1, z_2 \ldots z_n], \|I - \mathcal{Z}^T\mathcal{Z}\|/(2n\varepsilon) = 0.125$, actually the vectors associated with eigenvalues $-s_7$ and $-s_8$ span an eigenspace, resulting in singular vectors that are not orthogonal to work precision.

5   Numerical Experiments

We have used a large set of bidiagonal matrices to test BDSVDX, on a typical Intel-based computer, in double and single precisions, using different compilers. Here we report results in double precision only, with the gnu compiler. Most of the test matrices in our testbed are derived from symmetric tridiagonal matrices described in [2] (also used in [3]). In this case, we factor $T - \nu I = LL^T$ (Cholesky) for a proper value of ν (obtained from the Gerschgorin bounds of T), then set $B = L^T$. The testbed also includes bidiagonals generated with random entries. All matrices used in our experiments are available upon request.

To test accuracy, we compute $resid = \|U^T B V - S\|/(\|B\|n\varepsilon)$, $orthU = \|I - U^T U\|/(n\varepsilon)$, and $orthV = \|I - V^T V\|/(n\varepsilon)$, where n is dimension of B. To test RANGE = "I" or RANGE = "V" for a given B, we build the corresponding T_{GK} prior to invoking BDSVDX and compute its eigenvalues using bisection. Then, for RANGE = "V" we generate n_V pairs of random indices IL and IU, map those indices into the eigenvalues of T_{GK}, perturb the eigenvalues slightly to obtain corresponding pairs VL and VU, and then invoke BDSVDX n_V times. For RANGE = "I" we simply generate n_I pairs of random indices IL and IU, and then invoke BDSVDX n_I times.

Figure 1a shows the accuracy of BDSVDX, for all singular values and vectors, for 200 bidiagonal matrices with dimensions ranging from 9 to 4006. Figures 1b-c show the accuracy of BDSVDX for the same matrices of Fig. 1a, with $n_I = 10$ (random) pairs of IL, IU, and $n_V = 10$ (random) pairs of VL, VU for each matrix. In the figures, the matrices (y-axis) are ordered according to their condition numbers, which range from 1.0 to $>10^{200}$. For convenience, we use floor and ceiling functions to bound the results in the x-axis, setting its limits to 10^{-2} and 10^{+4}.

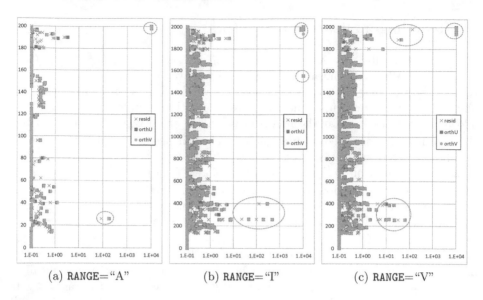

(a) RANGE="A"   (b) RANGE="I"   (c) RANGE="V"

Fig. 1. *resid, orthU, orthV* (*x*-axis, log scale) of BDSVDX for RANGE = "A","I" and "V", double precision. (1a) 200 matrices (*y*-axis), increasing condition numbers; (1b) $n_I = 10$ for each matrix of RANGE = "A"; (1c) $n_V = 10$ for each matrix of RANGE = "A".

As can be seen in Fig. 1a, the great majority of the results are adequately below 1.0. We consider the outliers to be the ones above 100 and mark them with an ellipsis. Matrix 26 is a bidiagonal matrix obtained from a tridiagonal matrix with highly clustered eigenvalues. Its dimension is 1260, its condition number is 2.2668, and its 136 largest eigenvalues have 12 digits in common (its spectrum contains other large clusters). Matrices 198–200 are more difficult: their entries are taken randomly from the interval $[2 * log(\varepsilon), -2 * log(\varepsilon)]$, therefore ranging from ε^{-2} to ε^2 (this is a notoriously hard case, borrowed from the LAPACK testers), and their dimensions are 125, 250 and 500, respectively. For $n = 500$, $s_1 = 1.47 \times 10^{+31}$ and $s_n = 1.34 \times 10^{-284}$ (as computed by BDSQR). For these matrices, resid is $O(10^{-8})$ but *orthU* and *orthV* are $O(10^{+13})$. As expected, the effect of large clusters of singular values of matrix 26 and the oddities of matrices 198–200, are propagated to Figs. 1b and c. Figure 1b contains additional outliers: case 398 corresponds to a bidiagonal similar to matrix 26 in Fig. 1a; cases 1551 and 1552 are related to a bidiagonal of dimension 1000, obtained from a tridiagonal with one eigenvalue equal to 1.0 and all others equal to $1/\sqrt{\varepsilon}$.

Figure 2 compares the times taken by BDSQR, BDSDC and BDSVDX on 12 bidiagonals with dimensions ranging from 494 to 2003 (a sample of matrices from Fig. 1a). For BDSVDX, we compute all singular values/vectors, the largest 20% and 10% singular values/vectors, and the largest 5 singular values/vectors. For each matrix, the timings are normalized with respect to the time taken by BDSQR (*y*-axis, log scale). As expected, BDSVDX is not competitive for all or a relatively large set of singular values/vectors, the gains become apparent at about 10%.

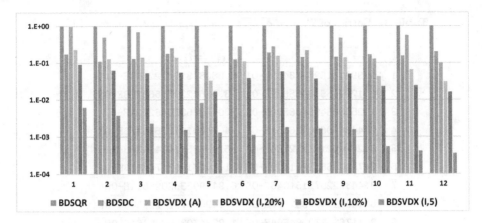

Fig. 2. Normalized times (y-axis, log scale) for BDSQR, BDSDC and BDSVDX on 12 bidiagonals whose dimensions range from 494 to 2003 (x-axis, increasing size), double precision. BDSVDX: all, the largest 20% and 10%, and the largest 5 singular values/vectors. For each matrix, the timings are normalized with respect to the time taken by BDSQR, which is typically the slowest.

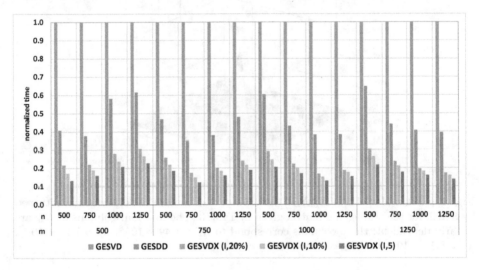

Fig. 3. Normalized times for GESVD, GESDD and GESVDX on 16 $m \times n$ matrices where $m = 500, 750, 1000, 1250$ and $n = 500, 750, 1000, 1250$, in double precision (real). GESVDX: the largest 20%, the largest 10% and the largest 5 singular values/vectors. For each matrix, the timings are normalized with respect to the time taken by GESVD.

In particular, BDSVDX is 3 orders of magnitude faster than BDSQR and 2 orders of magnitude faster than BDSDC for the computation of the largest 5 singular values and vectors of the largest matrix.

Table 3. Entries of T, $\lambda_i = c^{-\frac{(i-1)}{(n-1)}}, c = \frac{1}{\sqrt{\varepsilon}}, i = 1, 2, \ldots n, n = 10$.

i	$t_{i,i}$	$t_{i,i+1} = t_{i+1,i}$
1	1.893161597943482E-01	3.880873104122968E-01
2	8.128005558065539E-01	-3.516122075663728E-02
3	1.258328488738520E-01	3.077875339462724E-02
4	2.448430650126851E-02	-4.746410482563373E-03
5	3.268662131212184E-03	-6.983851144411338E-05
6	2.759036513821439E-04	-1.142712831766173E-04
7	9.443722972151846E-05	6.941905362025514E-06
8	6.149112437832172E-06	-7.426637317219540E-07
9	2.117627370984594E-07	1.892470326809461E-08
10	1.071603546505181E-07	-

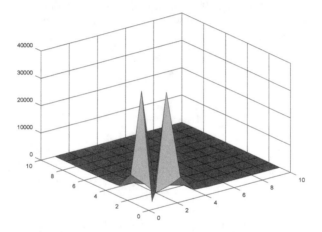

Fig. 4. Surface plot of $|I - X_{\text{XR}}^T X_{\text{XR}}|/(n\varepsilon)$, where X_{XR} contain the eigenvectors returned by STEXR for the tridiagonal matrix given in Table 3. The first four columns of X_{XR} are linearly dependent: those columns correspond to $\lambda_1 \approx 1.49 \times 10^{-8}$, $\lambda_2 \approx 1.10 \times 10^{-7}$, $\lambda_3 \approx 8.17 \times 10^{-7}$ and $\lambda_4 \approx 6.06 \times 10^{-6}$.

Finally, Fig. 3 compares the times taken by GESVD, GESDD and GESVDX in double precision, on random $m \times n$ matrices with $m = 500, 750, 1000, 1250$ and $n = 500, 750, 1000, 1250$. GESVDX is used to compute the largest 20%, the largest 10% and the largest 5 singular values/vectors. It is consistently faster than its counterparts, which are limited to a full SVD. The gains become more significant as the smallest dimension of the matrix increases: GESVDX is up to 7.5 times faster than GESVD and 2.8 times faster than GESDD.

A case of failure in STEXR. We show here a case of misbehavior of STEXR introduced in [6], by using a tridiagonal matrix T generated with the prescribed eigenvalue distribution $\lambda_i = c^{-\frac{(i-1)}{(n-1)}}, c = 1/\sqrt{\varepsilon}, i = 1, 2, \ldots n, n = 10$ ($\lambda_1 \approx$ $1.49 \times 10^{-8}, \ldots \lambda_n = 1.00$). Table 3 lists the entries of T. Although not shown here, the eigenvalues of T computed with the eigensolvers listed in Table 1 and also STEXR are in very good agreement. However, $\|I - X_{\text{XR}}^T X_{\text{XR}}\|/(n\varepsilon) \approx$ 3.95×10^4, where X_{XR} corresponds to the matrix of eigenvectors returned by STEXR; see Fig. 4. Noteworthy, $\|T - X_{\text{XR}}\Lambda_{\text{XR}}X_{\text{XR}}^T\|/(\|T\|n\varepsilon) \approx 0.6$, where Λ_{XR} is the diagonal matrix formed with the eigenvalues returned by STEXR. In contrast, $\|I - X_{\text{VX}}^T X_{\text{VX}}\|/(n\varepsilon) \approx 0.9$ and $\|I - X_{\text{MR}}^T X_{\text{MR}}\|/(n\varepsilon) \approx 0.1$, where X_{VX} and X_{MR} are the vectors returned by STEVX and STEMR, respectively. We have identified other matrices for which STEXR failed to produce orthogonal eigenvectors, for example the Wilkinson symmetric tridiagonal matrix of dimension 21, W_{21}, whose diagonal entries are $(10, 9, 8, \ldots, 0, \ldots 8, 9, 10)$ and whose offdiagonal entries are all 1. Our exhaustive tests revealed that STEMR may also fail for matrices with very close eigenvalues (e.g. matrices formed by gluing Wilkinson-type matrices). To the best of our knowledge, STEXR is not longer maintained, therefore our choice of STEVX for the first implementation of BDSVDX. We note that the computation of the bidiagonal SVD using MRRR has been also explored in [8] but the implementation discussed therein has not been incorporated into LAPACK; it has been phased out in favor of a more robust theory presented in [6,9].

6 Conclusions

This paper presented an algorithm for the computation of the SVD of a bidiagonal matrix by means of the eigenpairs of an associated tridiagonal matrix. The implementation, BDSVDX (included in the LAPACK 3.6.0 release), provides for the computation of a subset of singular values/vectors, which is important for many large dimensional problems that do not require the full set. Our experiments revealed that this feature can lead to impressive gains in computing times, when compared with existing implementations that are limited to the computation of the full SVD. The implementation discussed here offers opportunities for parallelism, for example by assigning different subsets of values and vectors to different processes. For a parallel implementation, we can built upon the workflow of the parallel subroutines PDSYEVX or PDSYEVR that are implemented in ScaLAPACK. The former is based on bisection and inverse iteration, with the caveat that it does guarantee orthogonality of eigenvectors that are on different processes. The latter is based on the MRRR algorithm and presumably delivers more satisfactory results and scalability [10]. Specific tests will be required (e.g. with cases similar to the difficult ones in Fig. 1) to assess the best alternative.

Numerical results on a large set of test matrices substantiated the accuracy of the implementation; the exceptions are matrices with very large condition numbers or highly clustered singular values. Interestingly, we have verified (results not shown) that the accuracy is not so much dependent on the condition number of the singular vectors, $\kappa_{u,v} = \min(\frac{1}{\min_i gap_i} \frac{1}{s_1}, \frac{1}{\varepsilon})$, $gap_i = \min_{j \neq i} |\sigma_i - \sigma_j|$, as we had originally thought. On the other hand, we have identified pathological cases (typically very small singular values) for which the computed singular vectors may not be orthogonal to work precision. A more robust strategy to cope with such cases needs to be investigated; it will be a priority in our future work.

References

1. Bai, Z., Demmel, J., Dongarra, J., Ruhe, A., van der Vorst, H. (eds.): Templates for the solution of Algebraic Eigenvalue Problems: A Practical Guide. SIAM, Philadelphia (2000)
2. Marques, O., Demmel, J., Voemel, C., Parlett, B.N.: A testing infrastructure for symmetric tridiagonal eigensolvers. ACM TOMS **35**, 1–13 (2008)
3. Demmel, J., Marques, O., Voemel, C., Parlett, B.N.: Performance and accuracy of LAPACK's symmetric tridiagonal eigensolvers. SIAM J. Sci. Comput. **30**, 1508–1526 (2008)
4. Dhillon, I.S., Parlett, B.N.: Multiple representations to compute orthogonal eigenvectors of symmetric tridiagonal matrices. Linear Algebra Appl. **387**, 1–28 (2004)
5. Dhillon, I.S., Parlett, B.N., Voemel, C.: The design and implementation of the MRRR algorithm. ACM TOMS **32**, 533–560 (2006)
6. Willems, P., Lang, B.: A framework for the MR^3 algorithm: theory and implementation. SIAM J. Sci. Comput. **35**, 740–766 (2013)
7. Demmel, J., Kahan, W.: Accurate singular values of bidiagonal matrices. SIAM J. Sci. Stat. Comput. **11**, 873–912 (1990)
8. Willems, P., Lang, B., Voemel, C.: Computing the bidiagonal SVD using multiple relatively robust representations. SIAM. J. Matrix Anal. Appl. **28**, 907–926 (2006)
9. Willems, P.: On MR^3-type algorithms for the tridiagonal symmetric eigenproblem and the bidiagonal SVD, PhD dissertation, University of Wuppertal (2010)
10. Voemel, C.: ScaLAPACK's MRRR algorithm. ACM TOMS **37**, 1–35 (2010)

HPC on the Intel Xeon Phi:
Homomorphic Word Searching

Paulo Martins[✉] and Leonel Sousa

INESC-ID, Instituto Superior Técnico, Universidade de Lisboa, Lisboa, Portugal
paulo.sergio@netcabo.pt, las@inesc-id.pt

Abstract. In this paper, the suitability of implementing parallel homomorphic word searching on Intel Xeon Phi coprocessors is evaluated for the first time. Homomorphic encryption allows to produce a cryptogram that encrypts the result of applying some values to any function, even when the input values are encrypted and without access to the private-key. For example, it is possible to search if any word of a set of encrypted words matches a plaintext reference word and generate a new cryptogram that encrypts the amount of matches. In this paper it is shown that this operation is about 834 times faster by using a system with 4 Intel Xeon Phi coprocessors 5110P attached to an Intel Xeon CPU E5-2630 v2, when compared with an implementation on a single core of the Xeon CPU.

Keywords: Intel Xeon Phi · Homomorphic encryption · Homomorphic word searching

1 Introduction

The use of embedded systems is becoming ubiquitous, as more sensors and actuators are incorporated into everyday electronics and on the general infrastructure. Since these devices often have limited computational resources, it would be beneficial to offload parts of their computation to a third party. However, the processed data may be private, which means that the third party should not have access to it. Cryptography enables the encryption of data, such that access to it is impossible without the usage of a specific key. In particular, with public-key cryptography, every user produces a pair of keys: one is public, and should be widely distributed, while the other is private. Someone with access to the public-key may produce a cryptogram by applying the encryption algorithm to a plaintext. This cryptogram cannot be decrypted by anyone but the owner of the corresponding private-key. Homomorphic Encryption (HE), in turn, allows one to operate on ciphered data [2]. With this approach, one can produce an encryption of the output of an arbitrary function from the encrypted inputs, and without access to the deciphering key.

P. Martins—This work was supported by national funds through FCT (Fundação para a Ciência e a Tecnologia) with reference UID/CEC/50021/2013 and under the Ph.D. grant with reference SFRH/BD/103791/2014.

© Springer International Publishing AG 2017
I. Dutra et al. (Eds.): VECPAR 2016, LNCS 10150, pp. 75–88, 2017.
DOI: 10.1007/978-3-319-61982-8_9

With homomorphic encryption, processors with limited resources may offload the processing of sensitive data without compromising the privacy of that data. This may be useful for instance, for advertising companies that wish to select ads based on the browsing history or keywords in e-mails of potential customers; or to analyze medical data of sensors connected to a patient without breaching the patient's privacy; or for companies wishing to offload the computation of the purchasing patterns of its customers. It is therefore of interest to investigate how current servers may execute these procedures in an efficient and scalable way. Some platforms to provide High Performance Computing (HPC) in server settings include various Field-Programmable Gate-Array (FPGA) solutions, the use of Graphics Processing Units (GPUs) for general purpose processing, among others. However, these solutions typically lead to high development and maintenance costs.

As an alternative, Intel developed the Many Integrated Core (MIC) architecture [5]. Intel MIC is a many core coprocessor architecture supported on a modified version of the P54C design, used on the original Pentium. These cores can be very power efficient on current semiconductor process architectures due to short pipelines and low frequency operations. Also, the modified cores enable the use of many of the programming models that most developers are already accustomed to, such as OpenMP, OpenCL, Message Parsing Interface (MPI), Cilk/Cilk Plus, and specialized versions of Fortran, C++ and math libraries. This is twofold important. First, there is a large amount of code already being deployed with these tools, that can be readily executed on the Intel Xeon Phi. Second, it eases the process of porting applications to the new architecture. We consider, therefore, that this architecture is one of the most suitable for servers with heterogeneous workloads, and investigate for the first time how well it is adapted to homomorphic word searching.

The rest of the paper is organized as follows. In Sect. 2, we give an introduction to HE, and describe how it can be applied to perform word searching. Afterwards, in Sect. 3, procedures and algorithms are proposed for the Intel Xeon Phi architecture. The performance of the proposed parallel algorithms is evaluated in Sect. 4, compared with related work in Sect. 5, and finally conclusions are drawn in Sect. 6.

2   Homomorphic Encryption

HE can be metaphorically explained by the jewelry shop problem [3], whose solution is represented in Fig. 1. Alice, a shop owner, wanted her workers to assemble precious materials, such as gold and diamonds, into intricately designed rings and necklaces. However, she distrusted her workers, and thus did not want the workers to come in direct contact with the materials, since she was afraid they might steal them. In order to solve this problem, Alice used a transparent impenetrable glovebox. She would then open the box, and store the raw materials inside. Afterwards, she would lock the box using a key to which only she had access. This process embodies the encryption procedure. As shown in Fig. 1,

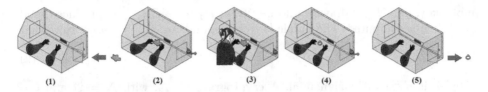

Fig. 1. A piece of gold is locked inside a glovebox, so that a worker may transform it into a ring. The ring is later removed when the glovebox is unlocked

the workers could use the gloves to assemble the rings and necklaces. In this situation, the gloves represent the homomorphism of an HE scheme. Since the box was impenetrable they could not have access to the precious materials; in a similar manner to how a server is not be able to access the encrypted values it processes homomorphically. When the piece was finished, Alice could open the box, and retrieve the result, which is mirrored as the decryption operation in the case of a cryptosystem.

Fully Homomorphic Encryption (FHE) was first uncovered in 2009 [2]. In contrast to previous Somewhat Homomorphic Encryption (SHE) schemes that only enabled a subset of all possible operations on cryptograms, with FHE it is possible to arbitrarily process encrypted data. In this paper, the cryptosystem described in [6] will be focused on. This cryptosystem is a leveled FHE scheme: with it, it is possible to evaluate arbitrary functions of encrypted data; one only needs to specify the maximum "size" of functions beforehand and the size of the generated public-key depends on this value. Arithmetic is performed in $R_q = \frac{\mathbb{Z}/(q\mathbb{Z})[x]}{\Phi(x)}$, with $\Phi(x) = x^n + 1$ and n a power of two. In this ring, two elements are congruent (i.e. equivalent) if their difference is a multiple of $\Phi(x)$. Similarly, two elements $a(x)$ and $b(x)$ are congruent if the difference between all corresponding coefficients a_i and b_i is a multiple k_i of q. An example of operations in this ring for $n = 2$ and $q = 2$ is as follows:

$$(x + 1) + x = 2x + 1 \equiv 1$$
$$(x + 1) \times x = x^2 + x \equiv x + 1 \tag{1}$$

where $\equiv$ is used to denote congruency. In the first case, if we consider the coefficients of x, one can see that $2 - 0 = 2$ is a multiple of $q = 2$, and therefore the congruency is valid. In the second equation, we can see that $x^2 - (-1)$ is a multiple of $\Phi(x) = x^2 + 1$, hence $x^2 + x \equiv x - 1$. Since $(-1) - 1$ is also a multiple of $q = 2$, the second congruency is valid in this ring. Typically, the elements of R_q with the smallest polynomial degrees, and with the smallest non-negative coefficients are used as the representatives for the congruency classes. With this representation, addition and multiplication of polynomials is followed by the computation of the remainder of the division by $\Phi(x)$, and afterwards by the computation of the remainder of the division of the coefficients by q.

In this cryptosystem, the secret-key corresponds to a vector $s_{2\times1} = (1, -t)^T \in R_q^2$, where $t \leftarrow D_{R_q, \sigma_k}$ is a polynomial drawn from a "narrow" distribution [6], namely a Gaussian distribution. In order to produce the corresponding public-key, $a \leftarrow R_q$ is drawn uniformly from R_q, $e \leftarrow D_{R_q, \sigma_k}$ is produced, and $b = at + e$ is computed. The public-key corresponds to $A_{1 \times 2} = (b, a)$. Note that

$$A_{1 \times 2} \times s_{2 \times 1} = e \tag{2}$$

where e is a "small" polynomial. A cryptogram $C_{N \times 2}$, with $N = 2l$ and $l = \lceil \log q \rceil$, encrypting a value $\mu \in R_q$ is a matrix such that, for a small $error$:

$$C_{N \times 2} \times s_{2 \times 1} = \mu \begin{bmatrix} 1 & 0 \\ 2 & 0 \\ \vdots & \vdots \\ 2^{l-1} & 0 \\ 0 & 1 \\ 0 & 2 \\ \vdots & \vdots \\ 0 & 2^{l-1} \end{bmatrix} \times s_{2 \times 1} + error \tag{3}$$

If we add two cryptograms $C_{N \times 2}$ and $D_{N \times 2}$, the resulting cryptogram retains the format described in Eq. (3). Homomorphic multiplication, in contrast, is more complex. To multiply two ciphertexts $C_{N \times 2}$ and $D_{N \times 2}$, one computes $BD(C_{N \times 2}) \times D_{N \times 2}$. The $BD(C_{N \times 2})$ function expands each entry of the matrix across l columns performing bit decomposition. In concrete, each element $c_{i,j}$ is decomposed into $c_{i,j}[k]$ for $k \in \{0, \ldots, l-1\}$, such that $c_{i,j} = \sum_{k=0}^{l-1} c_{i,j}[k] 2^k$, where the $c_{i,j}[k]$ are polynomials with coefficients either 0 or 1, producing a matrix:

$$\begin{pmatrix} c_{0,0}[0] & \cdots & c_{0,0}[l-1] & c_{0,1}[0] & \cdots & c_{0,1}[l-1] \\ c_{1,0}[0] & \cdots & c_{1,0}[l-1] & c_{1,1}[0] & \cdots & c_{1,1}[l-1] \\ \vdots & \ddots & \vdots & \vdots & \ddots & \vdots \\ c_{N-1,0}[0] & \cdots & c_{N-1,0}[l-1] & c_{N-1,1}[0] & \cdots & c_{N-1,1}[l-1] \end{pmatrix} \tag{4}$$

It can be proved that the result of $BD(C_{N \times 2}) \times D_{N \times 2}$ retains the format in (3), but μ now takes the value of the product of the original plaintexts. If the term $error$ remains small enough, the result can still be deciphered.

As more operations are performed, $error$ grows, and as such the number of operations that can be applied are limited by the homomorphic capacity of the cryptosystem. In the particular case of homomorphic multiplication, noise growth is asymmetric, i.e. if matrices $C_{N \times 2}$ and $D_{N \times 2}$ are swapped in the expression $BD(C_{N \times 2}) \times D_{N \times 2}$, the final $error$ will not necessarily be the same. It is best to use the ciphertext with the largest error as $C_{N \times 2}$ [6].

2.1   Ring Arithmetic

The parameters for the cryptosystem described in [6] enable efficient arithmetic over R_q. In particular, the value of n was set to $n = 1024$ and q to $q = \texttt{0x7FFE0001}$ in hexadecimal. This leads to $l = 31$ and $N = 62$. Reduction modulo q after multiplications is achieved by noting that the following congruence is valid (both side of the expression have the same remainder modulo q):

$$2^{31} \equiv 2^{17} - 1 \bmod q \tag{5}$$

since $q = 2^{31} - 2^{17} + 1$. Thus, the value $z \in \{0, \ldots, (q-1)^2\}$ of the product of two polynomial coefficients can be rewritten as $z = z_1 2^{31} + z_0$, and the following congruence can be applied:

$$z \equiv z_1 2^{17} + z_0 - z_1 \tag{6}$$

This latter congruence is iteratively employed as depicted in Algorithm 1 until $z \in \{0, \ldots, 2^{31} - 1\}$. Afterwards, a conditional subtraction by q when $z \in \{q, \ldots, 2^{31} - 1\}$ suffices to ensure that z is in $\{0, \ldots, q-1\}$. When adding or subtracting two polynomial coefficients, a subtraction or an addition by q suffices to bring the result z back to $\{0, \ldots, q-1\}$ when $z \geq q$ or $z < 0$, respectively.

Algorithm 1. Modular reduction in $\mathbb{Z}/(q\mathbb{Z})$

Require: $z \in \{0, \ldots, q^2 - 2q + 1\}$
Ensure: $z \in \{0, \ldots, q-1\}$
   while $z \geq 2^{31}$ **do**
      $z_1 = z >> 31$
      $z_0 = z \& (2^{31} - 1)$
      $z = z_1 2^{17} + (z_0 - z_1)$
   end while
   if $z \geq q$ **then**
      $z = z - q$
   end if
   return z

Addition of polynomials in R_q is performed by adding the corresponding polynomial coefficients in $\mathbb{Z}/(q\mathbb{Z})$ with Single Instruction Multiple Data (SIMD) instructions. Multiplication of two polynomials in $\frac{\mathbb{Z}/(q\mathbb{Z})[\dot{x}]}{\dot{x}^n - 1}$ is equivalent to a cyclic convolution of n points: if $u(\dot{x}) = \sum_{i=0}^{n-1} u_i \dot{x}^i$, $u(\dot{x})\dot{x}^k \equiv \sum_{i=0}^{n-1} u_{n-k+i \bmod n} \dot{x}^i \bmod \dot{x}^n - 1$, thus $z(\dot{x}) = u(\dot{x}) \times v(\dot{x}) \equiv \sum_{i=0}^{n-1} \sum_{j=0}^{n-1} u_{n-j+i \bmod n} v_j \dot{x}^i \bmod \dot{x}^n - 1$. This operation is equivalent to multiplying the coefficients of the Fast Fourier Transform (FFT) over $\mathbb{Z}/(q\mathbb{Z})$ of the two polynomials, which results in an algorithm with lower complexity. By noting that if $\eta^n \equiv -1 (\bmod q)$, and $\dot{x} = \eta x$, then

$$\dot{x}^n - 1 = (\eta x)^n - 1 \equiv -(x^n + 1) \equiv 0 (\bmod x^n + 1) \tag{7}$$

This means that if the change of variable $\dot{x} = \eta x$ is applied, operations modulo $\dot{x}^n - 1$ will be converted to operations modulo $x^n + 1$ when the variable is changed back to $x = \eta^{-1} \dot{x}$.

Thus, to multiply two polynomials $a(x)$ and $b(x)$ in R_q, one first computes $u(\dot{x}) \equiv \sum_{i=0}^{n-1} \underbrace{a_i \eta^{-i}}_{u_i} \dot{x}^i$ and $v(\dot{x}) \equiv \sum_{i=0}^{n-1} \underbrace{b_i \eta^{-i}}_{v_i} \dot{x}^i$. Afterwards, one applies a FFT over $\mathbb{Z}/(q\mathbb{Z})$ to u and v, and multiplies the resulting transforms coefficient-wise. To get the final result, an inverse FFT has to be applied, and a final change of variable to return the polynomials from $\dot{x}$ to x.

2.2   Homomorphic Word Matching

In this work, the previous scheme was applied to homomorphically perform word matching. One can imagine a server where e-mails are stored in encrypted format. The senders of e-mails should encrypt the words of those e-mails by applying Algorithm 2 to the set of words in the e-mail. It should be noted that since a hash function is used to conceal the words lengths it is not possible to obtain the plaintext words back from the cryptograms. For practical implementations, the sender would have to cipher the e-mail twice, once where all the words are encrypted with this algorithm, and another time where the e-mail is encrypted as a whole with a "reversible" encryption.

Algorithm 2. Encryption of a list of words to be searched

Require: List of words to be searched, *input_list*
Ensure: List of encrypted words, *output_list*
   output_list = {}
   for all *word* in *input_list* **do**
     encrypted_bit_list = {}
     a = Hash(*word*)
     for all bit a_i in a **do**
       c_i = Encrypt(a_i)
       encrypted_bit_list = *encrypted_bit_list* $\cup \{c_i\}$
     end for
     output_list = *output_list* $\cup \{$*encrypted_bit_list*$\}$
   end for
   return *output_list*

The e-mail client could then issue word searching queries. We assume, for simplicity, that the queried word is provided in the clear. The e-mail server would iterate through all encrypted words, and apply Algorithm 3 to each of them and the plaintext queried word. In this algorithm, one starts by computing the hash of the word to be searched, producing a. Afterwards, the value of Encrypt($\prod_i a_i$ XNOR b_i) is homomorphically computed, where b_i is the ith bit of the hash of the encrypted word. When processing the cryptograms, the term i in the product has the value of c_i if $a_i = 1$, and Encrypt(1) $- c_i$ otherwise, where c_i is the encryption of b_i. This operation produces a cryptogram that encrypts 1 when the two words are the same, or 0 otherwise. It should be noted that due to the asymmetric noise growth, it is best to compute the product linearly, instead of using a logarithmic tree. I.e., it is best to keep a product "accumulator" and multiply all terms by this accumulator sequentially. This allows one to always choose the accumulator of the product as the operand $C_{N \times 2}$ in $BD(C_{N \times 2}) \times D_{N \times 2}$, leading to a slower growth of the *error* term. Moreover, when one wants to perform a search over a set of encrypted words, one can apply Algorithm 3 to each of these words, and afterwards add the results to get the encrypted value of the number of matches. The server could then transfer this result back to the client, without ever having access to the amount of matches. The client could afterwards decipher the result.

Algorithm 3. Matching a plaintext word with an encrypted word

---

Require: Encrypted bits c_i of the hash of $word_1$
Require: Plaintext $word_2$
Ensure: Cryptogram *match* encrypts 1 if there was a match, and 0 otherwise
   $match = \text{Encrypt}(1)$
   $a = \text{Hash}(word_2)$
   for all bit a_i **in** a **do**
     if $a_i = 1$ **then**
       $d_i = c_i$
     else
       $d_i = \text{Encrypt}(1) - c_i$
     end if
     $match = BD(match) \times d_i$
   end for
   return *match*

In this work, only the more burdensome Algorithm 3 was parallelized and accelerated using Xeon Phis. The parallel implementation of this algorithm will be explained in detail in the following section.

3   Parallel Algorithms

The targeted system provided 4 Xeon Phi Knights Corner coprocessors [5]. Each coprocessor features 61 cores operating at 1.053 GHz, interconnected via a 512-bit bidirectional ring, as shown in Fig. 2. Since the Xeon Phi coprocessor runs an Operating System (OS) inside, one of the cores will typically be dedicated to answering hardware/software requests like interrupts. As such, there are 60 usable cores, each supporting four-way hyperthreading, and thus 240 hardware threads are available. The cores can run at turbo modes, increasing the frequency of operation, if the power envelope allows.

    Each core has two 32 kB L1 individual caches, for data and instructions, and a 512 kB L2 cache. The L2 caches are kept fully coherent by a global-distributed tag directory. The performance of the architecture is boosted with the vector

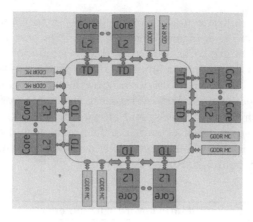

Fig. 2. Knights Corner schematic [11]

processing unit, which enables the processing of 512-bit registers. Each of the Xeon Phi cores has 32×512-bit SIMD registers. At each clock cycle, up to two instructions of a single thread are executed at each core. However, the two instructions follow two architecturally different pipelines, and therefore only one vector instruction can be executed at each cycle. The hardware cannot issue instructions back to back from the same thread in the core, and therefore at least two threads are necessary to reach full utilization of a core. Running 3 or 4 threads allows to hide more periods of latency, such as wrong instruction prefetches.

Since 512-bit SIMD instructions are available, 16 coefficients of a polynomial $f(x)$ can be processed in parallel, as each coefficient was represented with 32 bits. When performing reductions after additions or subtractions, comparison with q or 0 was implemented with the instruction vpcmpud, which produces a mask that indicates which lanes are greater or equal than q, or less than 0. This mask was used to prefix operations vpsubd and vpaddd that respectively subtract or add q to the lanes of the source register whose corresponding mask bit is 1. Furthermore, the repeated application of congruence (6) to reduce modular multiplications, as shown in Algorithm 1 was implemented with SIMD after unrolling the loop for when z initially had the value of $z = (q-1)^2$, so as to avoid divergent code on parallel operations.

Addition of polynomials in R_q was implemented by adding the corresponding polynomial coefficients over $\mathbb{Z}/(q\mathbb{Z})$ with SIMD instructions. Polynomial multiplications were implemented using changes of variable and FFTs. FFTs are decomposed into epochs, the number of which depends on the used radix r. In particular, each FFT consists of $\log_r n$ epochs, and in each epoch n/r computations, denominated butterflies, are performed. Using higher radices allows one to improve data locality. For the considered parameters, a radix-4 FFT was implemented, since $1024 = 4^5$. It is not possible to use higher radices (except for radix-1024, which is prohibitively large), since one cannot write 1024 as a power of a larger integer. Moreover, 16 butterflies were processed in parallel using SIMD extensions to speed up the FFT computation.

The homomorphic multiplication $E_{N \times 2} = BD(C_{N \times 2}) \times D_{N \times 2}$ proceeded in three steps:

- In step (i), multiple threads processed the $D_{N \times 2}$ matrix. It consists on changing the variable of the polynomials from x to $\dot{x}$, and afterwards applying the FFT, producing a new matrix $\hat{U}_{N \times 2} = FFT(CONV(C_{N \times 2}))$ (where $CONV$ denotes the change in variable).
- In step (ii), the matrix multiplication operation was processed in blocks. Each of the 240 threads in a Xeon Phi coprocessor was given an identifier-pair (id_x, id_y), with $id_x \in \{0, \ldots, 15\}$ and $id_y \in \{0, \ldots, 14\}$. By denoting $\hat{V}_{N \times N} = FFT(CONV(BD(C_{N \times 2})))$, $x_i = \lfloor \frac{id_x \times N}{16} \rfloor$, $x_f = \lfloor \frac{(id_x+1) \times N}{16} \rfloor$, $y_i = \lfloor \frac{id_y \times N}{15} \rfloor$, and $y_f = \lfloor \frac{(id_y+1) \times N}{15} \rfloor$, each thread performed the following operation:

$$\hat{W}^{(id_x, id_y)} = \begin{pmatrix} \hat{v}_{y_i, x_i} & \cdots & \hat{v}_{y_i, x_f - 1} \\ \vdots & \ddots & \vdots \\ \hat{v}_{y_f - 1, x_i} & \cdots & \hat{v}_{y_f - 1, x_f - 1} \end{pmatrix} \times \begin{pmatrix} \hat{u}_{x_i, 0} & \hat{u}_{x_i, 1} \\ \cdots & \cdots \\ \hat{u}_{x_f - 1, 0} & \hat{u}_{x_f - 1, 1} \end{pmatrix} \quad (8)$$

where the values of $\hat{v}_{i,j}$ are produced from the matrix $C_{N \times 2}$ as they are needed so as to reduce the memory requisites. Then, Algorithm 4 is used to add the blocks $\hat{W}^{(id_x, id_y)}$ with equal id_y to produce the matrix $\hat{W}_{N \times 2}$. In particular, this algorithm implements a logarithmic tree structure to add the intermediary results of the matrix: the base of the tree contains all intermediary values before running the algorithm, and the levels of the tree are processed from the base to the root. Each level corresponds to a time instant where the nodes are processed in parallel. In each of these nodes the values from its children are added, and therefore after the root is reached all intermediary results have been accumulated.
- In step (iii), an Inverse FFT (IFFT) is applied to the polynomials in $\hat{W}$, and $W = IFFT(\hat{W})$ is converted to $E_{N \times 2} = BD(C_{N \times 2}) \times D_{N \times 2}$, by changing the variable from $\dot{x}$ to x. Thread parallelism was used to process multiple polynomials simultaneously.

Since steps (i) and (iii) did not fully utilize the computational power of the Xeon Phi coprocessor, because there was not enough parallelism, several matrices were processed in parallel in these steps; which corresponds to searching on several words in parallel. Therefore, to compute s matrix multiplications, step (i) is first applied to s matrices in parallel, then step (ii) is repeated s times (once for each matrix multiplication), and finally step (iii) is applied s times in parallel to get the s results.

A modified version of Algorithm 3 was then implemented on the Xeon Phi. This modification corresponds to the comparison of the plaintext word with s encrypted words simultaneously, using the proposed matrix multiplication algorithm. Furthermore, the subtraction featured in the algorithm was accelerated

---

Algorithm 4. Logarithmic addition tree

Require: $\hat{W}^{(id_x,id_y)}, \forall id_x \in \{0,\ldots,15\}, \forall id_y \in \{0,\ldots,14\}$
Ensure: $\hat{W} = (\sum_{id_x} \hat{W}^{(id_x,0)}, \ldots, \sum_{id_x} \hat{W}^{(id_x,14)})$

$\hat{W}_{N \times 2} = 0$
The following code, until the return statement, is executed by all threads
for $hop = 2; hop \leq 16; hop = hop \times 2$ **do**
   Thread barrier
   if id_x is a **multiple of** hop **then**
     if $hop = 16$ **then**
       for $i = y_i; i < y_f; i = i + 1$ **do**
         $\hat{W}_{i,0} = \hat{W}^{(id_x,id_y)}_{i-y_i,0} + \hat{W}^{(id_x+hop/2,id_y)}_{i-y_i,0}$
         $\hat{W}_{i,1} = \hat{W}^{(id_x,id_y)}_{i-y_i,1} + \hat{W}^{(id_x+hop/2,id_y)}_{i-y_i,1}$
       end for
     else
       $\hat{W}^{(id_x,id_y)} = \hat{W}^{(id_x,id_y)} + \hat{W}^{(id_x+hop/2,id_y)}$
     end if
   end if
end for
Thread barrier
return $\hat{W}$

---

using multi-threading and SIMD extensions. When using a system with k Xeon Phis, the modified algorithm can be processed k times in parallel, and ks matches are homomorphically tested at the same time. Thus, when performing a search of a plaintext word over a set of encrypted words, the set was broken into smaller sets of ks encrypted words, and the Xeon Phis processed the sets in sequence. Afterwards, the host Central Processing Unit (CPU) adds the encrypted matches to create a cryptogram that when decrypted indicates how many encrypted words are equal to the plaintext.

4   Experimental Results

The proposed parallel algorithm was implemented on a system with an Intel Xeon CPU E5-2630 v2, operating at a frequency of 2.6 GHz, connected to 4 Intel Xeon Phi coprocessors 5110P, running at 1.053 GHz. The code was compiled with icc 16.0.1, using the optimization flag $-O2$. Computation was offloaded to the Xeon Phi coprocessors through icc pragmas, and the code on the Xeon Phi coprocessors was parallelized with OpenMP and SIMD intrinsics.

In order to find the optimal value for s (the number of matrix multiplications processed at a time by each Xeon Phi coprocessor), the homomorphic word searching algorithm was run on the 4 Xeon Phi coprocessors and timed for different values of s. In particular, a plaintext word was compared with sets of over 90 words for $s \in \{2, 4, 6, 8, 10, 12\}$. The relative execution time per encrypted word of the code offloaded to the Xeon Phi coprocessors can be found in Fig. 3. One can see that the relative search time per word decreases from 0.154 s for

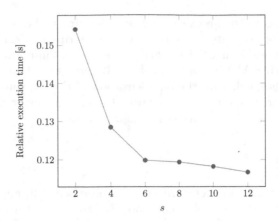

Fig. 3. Relative execution time per encrypted word as a function of the number of matrix multiplications (s) processed at a time by each Xeon Phi coprocessor

Table 1. Total execution time for homomorphic keyword searching

Number of encrypted words	Sequential execution [s]	Parallel execution [s]	Speedup
24	2662.6	3.6	743.9
48	5585.6	6.0	933.2
72	8418.6	10.5	803.2
96	11735	13.7	856.0

$s = 2$ until 0.119 s for $s = 6$, and stabilizes around that value for larger values of s. Therefore, the value of $s = 6$ was chosen for evaluating the performance of the parallel algorithms and obtaining the results presented next.

A sequential baseline version of Algorithm 3 was also implemented on a single core of the Xeon processor. Both the parallel implementation running on the Xeon Phi coprocessors and the sequential version running on the Xeon core were executed to perform a word search over sets of 24, 48, 72 and 96 encrypted words. The execution times of the word searching algorithm can be found in Table 1. There is a significant improvement in performance when executing the algorithm on the Xeon Phi coprocessors. In particular, an average speedup of 834 was obtained. By computing the obtained efficiency as the ratio between the speed-up and the product of the number of hardware threads and the number of SIMD lanes, and by taking into account the different frequencies of the Xeon processor and the Xeon Phi accelerator, one gets a value of 13.3%. This is accounted for by the fact that the vector unit is not exploited by all the instructions; that overhead is introduced when exploiting SIMD parallelism – for instance, in Algorithm 1 the while loop is unrolled for the worst case when exploiting SIMD extensions; and that there is also overhead associated with multi-threading parallelism – for example, when Algorithm 4 is executed, the underlying cache-coherence protocol introduces delay in every iteration, so that data sharing is possible.

Finally, the Intel Xeon CPU E5-2630 v2 processor supports up to 12 hardware threads, with SIMD instructions of 8×32-bits lanes. Hence, even if 100% efficiency were obtained, with the parallelization of the sequential Xeon homomorphic word searching operation, one would require a cluster 18 processors to beat the performance of the 4 Xeon Phi coprocessors.

5   Related Work

There has been work in the literature on how to port scientific applications to the Xeon Phi, such as the lattice Boltzmann code [13], the Monte Carlo tree search [9], the Rodinia benchmark [10] and sparse matrix multiplications [14] with very satisfactory performance. Common concerns among these works include the division of the work-load in a balanced way among the large amount of threads of the Xeon Phi coprocessor, an effective vectorization of code, and also of how to best distribute data in memory. While some works focus on the computation of the FFT [7,8], they are supported on the complex plane instead of finite fields, since they target telecommunications protocols, and hence their performance is not directly comparable with the FFT implementation presented in this work. Furthermore, long integer operations are optimized in [1] for the Xeon Phi coprocessor, with a special focus on vectorization. These operations are used to implement the Rivest-Shamir-Adleman (RSA) cryptosystem [12]. While textbook RSA is homomorphically multiplicative, since the multiplication of two cryptograms results in an encryption of the multiplication of the underlying plaintexts, textbook RSA is not considered safe, and this feature is not exploited in [1].

The described cryptosystem was supported on an earlier scheme [4], also based on matrix operations. However, the latter did not exploit ring arithmetic, which arguably degrades performance, and hence, as far as we know, there are no practical implementations. The cryptosystem proposed in [6] was also implemented therein using an Intel Core-i7 5930 K and a NVIDIA GeForce GTX980 as an accelerator. An homomorphic word searching procedure took a relative time of about 20 ms per encrypted word. This result cannot be directly compared with the results obtained herein (see Fig. 3) since the main objective of this work was to evaluate the improvement of performance one could get with widely deployed programming tools, such as OpenMP, that are available on the Xeon Phis, and provide more manageable code development. The GTX980 GPU is organized according to a different architecture, and targets a more strict range of applications, which does not allow a direct comparison with the results obtained for the Xeon Phi. Furthermore, the GTX980 GPU has 2048 CUDA cores, whereas the 4 Xeon Phis feature a total of 960 hardware threads, and therefore it is possible to exploit a larger level of parallelism with the GTX980. Also, the GTX980 runs at a slightly higher frequency (1.126 GHz) than the Xeon Phis (1.053 Ghz).

6   Conclusion

A large amount of embedded systems are currently being deployed in the market, either in the form of consumer electronics or household appliances and in the general infrastructure. Considering that they have limited computational resources, more and more computation will start being offloaded to central servers. Since this data may be private, it is expected that homomorphic encryption will become increasingly important, because it allows for the processing of encrypted data. In this work, the performance of homomorphic encryption is significantly enhanced with the use of Xeon Phi coprocessors. This enhancement is achieved by exploiting the fact that the considered cryptosystem relies on matrix multiplication over a specific ring, which is a burdensome operation with a large level of parallelism. In concrete, a speedup of about 834 was obtained for an homomorphic word searching procedure.

Furthermore, the Xeon Phi architecture has several advantages when compared with other HPC systems. It is more flexible than GPU architectures, supporting parallel divergent code more efficiently. It provides more manageable and less time consuming tools for code development than FPGAs. Finally, it supports multiple programming paradigms (such as OpenMP, and MPI) that are widely deployed, and for which large codebases already exist.

References

1. Chang, C., Yao, S., Yu, D.: Vectorized big integer operations for cryptosystems on the Intel MIC architecture. In: 2015 IEEE 22nd International Conference on High Performance Computing (HiPC), pp. 194–203, December 2015
2. Gentry, C.: A fully homomorphic encryption scheme. Ph.D. thesis, Stanford University, Stanford, CA, USA (2009)
3. Gentry, C.: Computing arbitrary functions of encrypted data. Commun. ACM **53**(3), 97–105 (2010). http://doi.acm.org/10.1145/1666420.1666444
4. Gentry, C., Sahai, A., Waters, B.: Homomorphic encryption from learning with errors: Conceptually-simpler, asymptotically-faster, attribute-based. Cryptology ePrint Archive, Report 2013/340 (2013). http://eprint.iacr.org/2013/340
5. Jeffers, J., Reinders, J.: Intel Xeon Phi Coprocessor High Performance Programming, 1st edn. Morgan Kaufmann Publishers Inc., San Francisco (2013)
6. Khedr, A., Gulak, G., Vaikuntanathan, V.: Shield: scalable homomorphic implementation of encrypted data-classifiers. Cryptology ePrint Archive, Report 2014/838 (2014). http://eprint.iacr.org/
7. Khelifi, M., Massicotte, D., Savaria, Y.: Parallel independent FFT implementation on intel processors and Xeon phi for LTE and OFDM systems. In: Nordic Circuits and Systems Conference (NORCAS): NORCHIP International Symposium on System-on-Chip (SoC), pp. 1–4, October 2015
8. Khelifi, M., Massicotte, D., Savaria, Y.: Towards efficient and concurrent FFTs implementation on Intel Xeon/MIC clusters for LTE and HPC. In: 2016 IEEE International Symposium on Circuits and Systems (ISCAS), pp. 2611–2614, May 2016

9. Mirsoleimani, S.A., Plaat, A., Van Den Herik, J., Vermaseren, J.: Scaling Monte Carlo tree search on Intel Xeon phi. In: 21st International Conference on Parallel and Distributed Systems (ICPADS), 2015, pp. 666–673, December 2015
10. Misra, G., Kurkure, N., Das, A., Valmiki, M., Das, S., Gupta, A.: Evaluation of rodinia codes on Intel Xeon Phi. In: 2013 4th International Conference on Intelligent Systems, Modelling and Simulation, pp. 415–419, January 2013
11. Reinders, J.: An overview of programming for Intel Xeon processors and Intel Xeon Phi coprocessors, November 2012. https://software.intel.com/en-us/mic-developer, Intel Corporation
12. Rivest, R.L., Shamir, A., Adleman, L.: A method for obtaining digital signatures and public-key cryptosystems. Commun. ACM **21**(2), 120–126 (1978). http://doi.acm.org/10.1145/359340.359342
13. Rosales, C.: Porting to the Intel Xeon Phi: opportunities and challenges. In: 2013 Extreme Scaling Workshop (XSW 2013), pp. 1–7, August 2013
14. Saule, E., Kaya, K., Çatalyürek, Ü.V.: Performance evaluation of sparse matrix multiplication kernels on Intel Xeon Phi. In: Wyrzykowski, R., Dongarra, J., Karczewski, K., Waśniewski, J. (eds.) PPAM 2013. LNCS, vol. 8384, pp. 559–570. Springer, Heidelberg (2014). doi:10.1007/978-3-642-55224-3_52

A Data Parallel Algorithm
for Seismic Raytracing

Allen D. Malony[✉], Stephanie McCumsey, Joseph Byrnes, Craig Rasmusen,
Soren Rasmusen, Erik Keever, and Doug Toomey

University of Oregon, Eugene, USA
malony@cs.uoregon.edu

Abstract. Dijkstra's single-source shortest path algorithm has been
applied in seismic tomography to determine paths of minimum travel
time from all locations in a 3D earth model to sensors used in seismic
experiments. An iterative data parallel algorithm is formulated for seismic tomography based on the Bellman-Ford-Moore (BFM) algorithm.
Performance is demonstrated for OpenMP on multicore processors and
OpenCL on GPUs.

Keywords: Seismic tomography · Shortest path · Data parallel

1 Introduction

A common problem in scientific computing is finding the shortest path from a
point to all other points in a given dataset. One of the most commonly used algorithms was first described by Dijkstra in his famous 1959 paper, "A Note on Two
Problems in Connexion with Graphs" [4]. Given a graph of n nodes each with
cost u and a starting node s, Dijkstra's "single-source shortest path" (SSSP)
algorithm finds the path from s to all other nodes with minimum cost. Unfortunately, Dijkstra's algorithm is difficult to implement in parallel. In this study, we
describe a data parallel algorithm for finding shortest paths on a regularly-spaced
grid of points that can compete with Dijkstra's in seismic raytracing.

Just as doctors use x-ray tomography to image the internal structure of the
human body, scientists studying the Earth use seismic tomography to image
its interior. Seismic waves propagate through the Earth at a velocity that varies
with local temperature, composition, and the presence of magma. Understanding
how these factors vary within the Earth is crucial to understanding the dynamic
processes that shape the planet. The method works by measuring the arrival
times of seismic waves from a source of seismic radiation, often an earthquake
or an explosion, and comparing the observed arrival times to the arrival times
predicted with a starting model. Perturbations to the starting model are then
solved by minimizing the misfit between the predicted and measured arrivals
times, generally with one of several variations on a least-squares approach [10].

In many cases, perturbations to the starting model are large enough to change
the geometric ray paths of the first-arriving waves [15]. Ray paths for the new

© Springer International Publishing AG 2017
I. Dutra et al. (Eds.): VECPAR 2016, LNCS 10150, pp. 89–98, 2017.
DOI: 10.1007/978-3-319-61982-8_10

model must be calculated before new perturbations to the model are calculated. This process increases the computation time of the algorithm from minutes to many hours. For example, Bezada et al. [2] implemented Dijkstra's algorithm for a tomographic study of the upper mantle beneath Spain and Morocco. While the improved results settled a long-standing controversy regarding the tectonic history of the Western Mediterranean Ocean, Dijkstra's algorithm had to be run for 322 starting points over 8 iterations. A significant amount of computing time was used for this purpose.

In general, reducing computation time is necessary to improve seismic tomography for several reasons. Seismic tomography is an ill-posed inverse problem, and several subjective parameters must be chosen to define the inverse problem. Since these parameters influence the final model, many inversions employing different parameter values must be analyzed before reliable inferences about structures inside the Earth can be made. The resolving power of the available data must also be explored, often through the inversion of many synthetic data sets. Finally, the computation time taken by Dijkstra's algorithm limits the scale of the problems that can be addressed. Efficiency of the shortest path problem must be improved for application to modern tomographic algorithms.

In this study, we present an alternative algorithm to that of Dijkstra [4] that is more amenable to parallelization. The algorithm's origin goes back to the famous Bellman-Ford-Moore (BFM) [1,6,8] algorithms which iterates to determine shortest paths. In contrast to Dijkstra's direct solution, ours then is an indirect computation that must iterate to reach a converged state. Nevertheless, the significant increase in parallelism enabled by the algorithm translates to overall reductions in execution times, as demonstrated for applications written with OpenMP and OpenCL. More importantly, the algorithm will hopefully expand the scale of the problems that can be addressed with seismic tomography and aid in the rapid development of high quality models of seismic velocity.

2   Stingray

Moser's formulation of the Dijkstra's shortest-path (DSP) method for seismic tomography [9] as implemented by Toomey et al. [15] is referred to in this paper as *Stingray-DSP*. Moser's approach represents the seismic velocity of a region of the Earth by a 3D grid of $N = N_x \times N_y \times N_z$ points. The objective is to find the shortest path from a starting point s to all other points p in the grid. This is done by initializing the travel time to s at all points p to ∞, and the travel time at point s to zero. The travel time to all the points

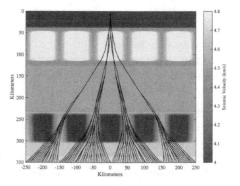

Fig. 1. Visualization of ray paths in a 2D velocity model.

in a neighborhood near s are then calculated along straight line paths. (The travel time between any 2 points p and q is a function of the velocity values at p and q and the distance between them.) The neighborhood of points used is called the "forward star" for s, $FS(s)$. In this first step, the infinite travel times for all the points in $FS(s)$ are replaced by the travel time from s.

Once the travel time to all of the points within $FS(s)$ are found, the point p with the minimum travel time from s is identified. The algorithm moves to p and then the travel times for $FS(p)$ are found. This is the standard DSP step. Again, only the minimum travel time for each point in $FS(p)$ is kept. Travel times along paths that are not minimum time are discarded. The algorithm continues with the next point q with minimum travel time which has not yet had a forward star centered on it. The process is repeated until a forward star had been centered at every point in the velocity model.

The accuracy of the Moser method primarily depends on how finely the model is discretized and how many points are included in the forward star. The error varies with both these quantities because both control how the angles of the ray paths leaving the center of the forward star are discretized. In pratice, we generally use a forward star of radius 7^1, with certain points removed that do not effect the final accuracy of the solution [7]. In practice, a total of 818 points are included in the reduced forward star of size 7. Figure 1 shows an example of the fidelity of ray paths that can be obtained with Stingray-DSP with an 818-point forward star resolution.

Various modifications to the algorithm are often to made to accommodate the scientific problem being addressed. Seismic velocities arc often anisotropic, that is, the seismic velocity depends not only on local properties of the rock but on the direction of the ray path. Three extra arrays must be stored in addition to the seismic velocity (the fractional magnitude of anisotropy, and the dip and azimuth of the fast direction) to compute anisotropic velocities, but the algorithm is essentially unchanged. Many starting points can be initialized at once to study the radiation of waves from an interface or a plane wave instead of a point source. Essentially, the solution for each starting point becomes a separate run of the algorithm. Finally, the choice of how the velocities are calculated along a ray path can be varied based on the complexity of the problem. Often, the velocity along a single ray is found by averaging the velocity at the end points. However, if velocities vary on a length scale shorter than the size of the forward star, a more expensive approach of integrating the velocities along the ray can be used. The run times given here all use end point averaging, which is appropriate for most geologic applications [2].

---

[1] A forward star around point p of radius r will include all grid points within a distance of $\delta * r$ from p, where δ is the grid point unit spacing. Some of those points will be redundant (e.g., colinear points) and can be removed from consideraton.

3 Stingray Iterative Constraint Convergence Algorithm

The Stingray-DSP algorithm gives seismic modeling scientists high-quality ray-tracing results compared to other methods. However, there are inherent limitations on parallelism in the algorithm that prevent high-performance computing (HPC) implementations. At each step of the algorithm, it is necessary to find the leaf point on the unfolding spanning tree that has the minimum travel time. This point must be the next one to expand, effectively sequentializing the control path. It is possible to execute the Stingray-DSP algorithm on multiple starting points at the same time, thus taking advantage of multiple computing resources. Over a hundred starting points for a single velocity model are used routinely in our work. Replicated parallelism is beneficial to geological scientists for through-put purposes, but it does not produce a faster single starting point solution. Also, very large velocity models could exceed the memory bounds of a single process-ing node, requiring a splitting of the Stingray-DSP computation across nodes. In general, DSP algorithms face fundamental performance inefficiencies when executing in distributed memory systems.

It is possible to reformulate seismic raytracing as an *iterative constraint convergence* (ICC) problem, where the constraint is the minimization of a travel time metric. Let V be the velocity field defined on a 3D grid of points and T the travel times from each grid point to a starting point s. Assuming the final travel times are known for each model point p, $T(p)$, they must satisfy the constraint:

$$T(p) = min(T(q) + Delay(p,q)), \ \forall \, q \in FS(p) \tag{1}$$

where $FS(p)$ is the "forward star" set of points of p and $Delay(p,q)$ is the seismic time delay (determined by the velocity values, plus additional physcial properties, in the case of anisotropic analysis) at p and q. The reason is that the minimum travel time path from p to s must pass through a point, r, in $FS(p)$, and r must be the point in $FS(p)$ whose own travel time to s, plus the delay from p to r is the smallest. Any other point can not be on the minimum travel time path from p to s. Based on this final constraint, an interative procedure to update the travel times can be specified as follows:

$$T_0(p) = \infty \ \forall \, p \neq s, \ T_0(s) = 0 \tag{2}$$

$$T_{i+1}(p) = min(T_i(q) + Delay(p,q)) \ \forall \, q \in FS(p) \tag{3}$$

where $T_i(p)$ and $T_{i+1}(p)$ are the travel times of p to s at steps i and $i+1$, respec-tively. The procedure continues until $T_{i+1}(p) = T_i(p) \ \forall \, p$. Note, convergence is guaranteed because the travel times at each point are monotonically decreasing.

The Stingray-ICC algorithm formulated in this manner is highly data-parallel, in that all points can be updated simultaneously. However, time to solution will depend on how long the algorithm takes to converge. There are three issues to consider. First, at step 0, all points have a time of ∞, except for the starting point. Thus, much of the early computation will be irrelevant and wasted until valid travel times radiate from the starting point. Second, the prop-agation of valid travel times is directly correlated with the radius of the forward star. Geological scientists prefer forward stars with larger radii for better accuracy, which will radiate travel times faster and hopefully result in fewer steps for convergence, but will also increase the computational work at each step. Third, as the iterative algorithm gets closer to convergence, fewer travel times will be adjusted, meaning more points will be already at their final travel times and the computation will be redundant.

4 Parallelization Design Strategy

The highly-parallel nature of the Stingray-ICC algorithm provides an excellent opportunity for parallelization on both multicore CPUs and manycore coprocessors. Ideally, we would like to articulate a parallelization design model that could map to different execution targets. The idea is to decompose the model domain into rectangular regions that can be worked on in parallel at each iteration step. The regions will be defined such that they are non-overlapping, in order to eliminate dependencies between regions during the step-wise parallel computation. However, between steps, exchanges between neighbor regions will be required to update the travel times for points on the region boundaries. This is a standard domain decomposition approach with halos used for exchanging boundary data. Typically, applications using domain decomposition will apply stencils in updating values within a region. The forward star in Stingray-ICC is effectively a stencil. The problem is that the 818-point forward star is a very large stencil. This makes it more challenging.

In order to update the travel time of a single point, a region in the Stingray-ICC domain decomposition must be at least of size $15 \times 15 \times 15$ in order to hold all of the points in the 818-point forward star. For a $150 \times 150 \times 150$ velocity model, this partitioning would generate 1000 regions. Once the partitioning is done, the objective is to update every point in the region in parallel across all regions, at each step. However, to do so, we would need to access the forward star around every point in the region. That requires information to be exchanged with our region neighbors to get those forward star points that are outside the region boundary. (Only, the point in the center of the region has it entire forward star set of points contained in the region.)

Deciding on halo size is essentially a tradeoff of extra buffer space versus when the exchanges with neighbors must be made. To accommodate all points in neighboring regions needed to update all points in a $15 \times 15 \times 15$ region with a 818-point forward star, a *region + halo* dimension of $22 \times 22 \times 22$ is necessary. Figure 2 illustrates the decomposition approach. It shows how the forward star defines the boundary overlap and the resulting halo surrounding the region.

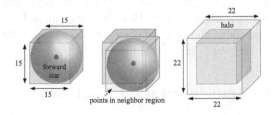

Fig. 2. Illustration of $15 \times 15 \times 15$ region, forward star, and $22 \times 22 \times 22$ region with halo.

The general parallelization design strategy above provides a basis for translation to target environments. In doing so, there are some additional strategies we can apply. For instance, in multicore shared memory systems, where multithreading is used to process regions (1 thread per region), it is possible to avoid the allocation of halos altogether by scheduling which region points are updated when in a cooperative manner with neighbor regions. The basic idea is illustrated in Fig. 3. Inspired by alternating direction implicit methods [5], the top row shows how points in a region could be processed in sweeps across the X (left), Y (middle), and Z (right) directions. (Reverse sweeps are also shown.) By coordinating neighbor regions in synchronous sweeps, forward star points in neighbor regions can be accessed directly without memory races. This is shown in the bottom row for two neighbor regions in the X, Y, and Z orientations. The strategy above could also have benefit in translation to manycore coprocessors, but more specialization will likely be required, especially for GPU accelerators.

A strategy to improve convergence is made possible by a slight addition in the Stingray-ICC algorithm. At every step, the algorithm updates the travel time of a point p by checking the travel times and delays of the points in its forward star. In doing so, the following condition might occur:

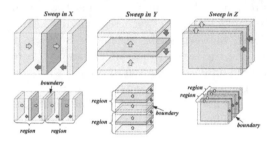

Fig. 3. Illustration of sweep methods for coordinated scheduling in and between regions.

$$T_i(q) \; < \; T_i(p) \; + \; Delay(p, q) \tag{4}$$

This means that we have discovered a better travel time for q. The strategy then is to update q's travel time opportunistically:

$$T_i'(q) \; = \; T_i(p) \; + \; Delay(p, q)) \;\; if \; T_i(q) \; < \; T_i(p) \; + \; Delay(p, q) \tag{5}$$

The notation $T_i'(q)$ is used to indicate that the update occurs in step i. The intuition is that any travel time updates carry new information, potentially improving convergence rate. However, care must be taken with this strategy to ensure that new memory race conditions are not introduced. Combining it with the "sweeping" strategy above will help.

5   Implementation Approach

Our objective was to compare the Stingray-DSP implementation of Moser's method with different implementations of the Stingray-ICC algorithm. The Fortran Stingray-DSP code runs sequentially for a velocity model and single starting point. Travel times for multiple starting points can be solved by replicating the Stingray-DSP execution across computing threads.

The Stingray-ICC algorithm was implemented for both a CPU and GPU. The CPU code was written in C with OpenMP for parallelization. The *Stingray-ICC-multistart* version will execute the algorithm sequentially, but for multiple starting points. This provides a close approximation to how the Stingray-DSP program is used in practice. The *Stingray-ICC-parsingle* parallelizes the algorithm for a single starting point. The *Stingray-ICC-gpu* program was adapted from the original Fortran source using CoArray Fortran extensions (CAFe) to communicate with and run OpenCL kernels on the GPU. CAFe allows the programmer to explicitly allocate memory on the GPU, transfer memory between the CPU and the GPU, and execute OpenCL kernels using coarray Fortran [12,14] syntax. CAFe is implemented as an embedded Domain Specific Language (DSL) and CAFe source is transformed automatically to standard Fortran [11], with wrappers [13] implementing the OpenCL C library interfaces. The OpenCL kernels implementing the *Stingray-ICC-gpu* algorithm were coded by hand.

6 Experimental Results

To evaluate the performance and scaling behavior of the Stingray-DSP and Stingray-ICC codes, we ran a series of experiments on different velocity models and sizes. These are described in Table 1. The *v100*, *v150*, *v200*, and *v300* models are synthetically generated by chosing a velocity value randomly within a velocity range for each model

Table 1. Velocity model descriptions.

Model	X Dim	Y Dim	Z Dim	# Points
v100	100	100	100	1000000
v150	150	150	150	3375000
v200	200	200	200	8000000
v300	300	300	300	27000000
v241	241	241	51	2962131

point. The *v241* model is taken from a real-world example. Each model is run with 12 starting points. This is done in Stingray-DSP by replicating the code as a separate process on each core of the CPU server. This is done in the Stingray-ICC-multistart code with OpenMP. An additional set of experiments using the *v241* model and a single starting point were conducted with the Stingray-ICC-parsingle for 1, 2, 4, 8, and 12 threads.

The shared memory machine used for our study was a HP ProLiant SL390 G7 server with two Intel X5650 2.66 GHz 6-core CPUs (12 cores total) and 72 GB DDR3 memory. Two GPUs were used: a NVIDIA M2070 (448 CUDA cores, 6 GB) and NVIDIA K80 (2496 CUDA cores, 12 GB).

Figure 4 (left) shows how the performance scales for the synthetic models and different codes. The Stingray-ICC versions perform significantly better than Stingray-DSP. Both Stingray-DSP and Stingray-ICC-multistart solve for 12 starting points, where each is run sequentially on 1 of 12 cores. Thus, these times reflect how long a serial execution for 1 starting point would take. In contrast, the Stingray-ICC-gpu results also solve for 12 starting points, but one after the other. We plot the average execution time for a single starting point for each

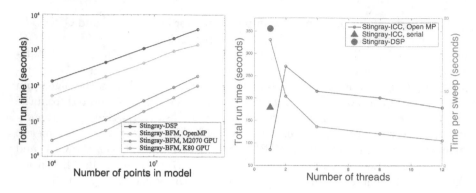

Fig. 4. Performance with synthetic and real velocity models.

GPU. Note, the average number of steps to reach convergence are constant at 6 steps for Stingray-ICC-multistart, but increase from 21 (v100) to 61 (v300) for Stingray-ICC-gpu.

The story gets more intriguing moving to the *v241* model experiments. Figure 4 (right) shows results from running Stingray-DSP and Stingray-ICC-multistart on 12 starting points. Again, Stingray-ICC-multistart is faster and it takes 7 steps to reach convergence for all 12 starting points. Figure 4 (right) also plots Stingray-ICC-parsingle results for 1, 2, 4, 8, and 12 threads, run with a single starting point. In this case, only 5 steps are needed to converge for 1 thread. However, the convergence steps increase from 2 to 12 threads (29 to 53 steps), though the time per sweep improves from 17.02 (1 thread) to 3.37 (12 threads). The increase in convergence steps nullifies the parallel performance gains (12 threads take 178.4 seconds). Note, the GPU times for the *v241* model were less than 10 secionds.

7   Discussion

Dijkstra's algorithm in Stingray-DSP only visits each point in the model once. Thus, the number of steps is determined by the number of points N in the model. In contrast, the ICC algorithm visits every vertex in each sweep of the model until the solution converges. Thus, the ICC execution time will be determined by the time per iteration multiplied by the number of iterations necessary for convergence. While the Stingray-ICC implementations are running faster than the Stingray-DSP code we have used for many years (which is certainly a welcome surprise), we notice that the number of convergence steps increases with larger problem sizes and more parallelism. Our goal is to scale to much larger seismic tomography problems with data parallel methods like ICC. If we can not get the convergence better under control, scaling limits might occur.

There is an interesting tradeoff in parallelism and convergence. We see the time per iteration decreasing in the *v241* model experiments with Stingray-ICC-parsingle. However, we believe the convergence steps increase because the sweep algorithm becomes more localized for each core and therefore less effective in propagating knowledge about shortest delay paths to its neighbors. Performance will improve with greater degrees of parallelism as long as the per iteration time reduces fast enough to offset more convergence steps. From the trajectory of the graph, we believe that great numbers of cores (e.g., as on the Xeon Phi) will allow OpenMP to obtain faster execution times.

Clearly, the Stingray-ICC-gpu execution times on the two GPUs (NVIDIA M2070 and K80) are taking significant advantage of data parallelism. The increase in the number of CUDA cores in the K80 also demonstrates the benefit of greater parallelism. The new NVIDIA Pascal architecture should deliver even faster execution.

In general, the ICC algorithm as implemented in this study is ignorant of anything having to do with the seismic model and the starting point. In fact, where the starting point is located does affect the convergence rate. In contrast, Stingray-DSP begins at the starting point. We believe that the runtime of the ICC algorithms can be improved by considering the behavior of the DSP "wavefront" propagation. Starting at the source, the wavefront will expand in roughly an oblong shape with deviations from a sphere due to anisotropies in the velocity model. Dijkstra's algorithm calculates the travel time from the starting point to its nearest neighbor (in time), then calculates the next nearest neighbor, and so on. At any given travel time, the set of vertices updated with this travel time will approximately map out the oblong shape of the expanding wavefront. If we can approximate this type of wavefront in how the ICC algorithm deicides which point to process, convergence rates might improve. This is currently being investigated.

8   Related Work

Methods for parallelizing Dijkstra's SSSP have been developed and recent work targets GPU implementations [3]. However, these have not been used the field of seismic tomography to solve the problems we consider here. Recasting the DSP approach to seismic raytracing as an iterative constraint convergence algorithm for parallelization purposes is similar to what is being done in calculating accumulated cost surfaces (ACS) [16] in spatial modeling. The BFM algorithm is the fundamental basis for both, except ACS applications are typically in 2D, such as in spatial analysis of raster images to determine route travel times. Speedup on ACS problems has been demonstrated with the BFM-inspired data parallel algorithm when targeting GPU.

9   Conclusion

Geological scientists turn to seismic raytracing as a preferred solution to create high-resolution tomographic models of the earth's interior. However, seismic raytracing based on Dijkstra's "single-source shortest path" (SSSP) algorithm can not take full advantage of parallel computing. We have described and demonstrated an alternative algorithm for seismic raytracing by reformulating the problem as an iterative constraint convergence algorithm. The Stingray-ICC approach is more amenable to parallelization and hence significantly reduces the computation time needed to calculate high quality seismic velocity models. We have demonstrated the application of the algorithm with OpenMP and OpenCL for GPUs. The use of this algorithm in the future will aid seismologists in enhancing our understanding the internal structure and dynamic behavior of our ever mysterious planet.

References

1. Bellman, R.: On a routing problem. Q. Appl. Math. **16**, 87–90 (1958)
2. Bezada, M., Humphreys, E., Toomey, D., Harnafi, M., Davila, J., Gallart, J.: Evidence for slab rollback in westernmost mediterranean from improved upper mantle imaging. Earth Planet. Sci. Lett. **368**, 51–60 (2013)
3. Davidson, A., Baxter, S., Garland, M., Owens, J.: Work-efficient parallel GPU methods for single-source shortest paths. In: International Parallel and Distributed Processing Symposium, pp. 349–359. IEEE, May 2014
4. Dijkstra, E.: A note on two problems in connection with graphs. Numer. Math. **1**, 269–271 (1959)
5. Douglas, J.: Alternating direction methods for three space variables. Numerische Mathematik **4**(1), 41–63 (1962)
6. Ford, L.: Network Flow Theory. RAND Corporation (1956)
7. Klimes, L., Kvasnicka, M.: 3-D network ray tracing. Geophys. J. Int. **116**(3), 726–738 (1994)
8. Moore, E.: The shortest path through a maze. In: International Symposium Switching Theory, pp. 285–292. Harvard University Press (1957)
9. Moser, T.: Shortest path calculation of seismic rays. Geophysics **56**(1), 59–67 (1991)
10. Nolet, G.: A Breviary of Seismic Tomography: Imaging the Interior of the Earth and Sun. Cambridge University Press, New York (2008)
11. Rasmussen, C., Sottile, M., Rasmussen, S., Nagle, D., Dumars, W.: Cafe: coarray fortran extensions for heterogeneous computing. In: 21st International Workshop High-Level Parallel Programming Models and Supportive Environments, HIPS 2016 Chicago, IL, USA, 23 May 23, 2016, Proceedings (2016)
12. Reid, J.: The new features of fortran 2008. SIGPLAN Fortran Forum **27**(2), 8–21 (2008)
13. Sottile, M., Rasmussen, C., Weseloh, W., Robey, R., Quinlan, D., Overbey, J.: ForOpenCL: transformations exploiting array syntax in fortran for accelerator programming. Int. J. Comput. Sci. Eng. **8**(1), 47–57 (2013)
14. The Fortran Committee. TS 18508 Additional parallel features in Fortran. ISO/IEC JTC1/SC22/WG5 N2007, March 2014
15. Toomey, D., Solomon, S., Purdy, G.: Tomographic imaging of the shallow crustal structure of the East Pacific Rise at $9°30'$. J. Geophys. Res. **99**, 24–24 (1994)
16. Trunfio, G., Sirakoulis, G.: Computing multiple accumulated cost surfaces with graphics processing units. In: International Conference on Parallel, Distributed, and Network-based Processing (PDP). Euromicro (2016)

Performance Modeling and Analysis

A Cross-Core Performance Model
for Heterogeneous Many-Core Architectures

Rui Pinheiro, Nuno Roma, and Pedro Tomás[(✉)]

INESC-ID, Instituto Superior Técnico, Universidade de Lisboa, Lisbon, Portugal
pedro.tomas@inesc-id.pt

Abstract. An accurate performance predictor to identify the most suitable core-architecture to execute each thread/workload in a heterogeneous many-core structure is proposed. The devised predictor is based on a linear regression model that considers several different parameters of the many-core processor architectures, including the cache size, issue-width, re-order buffer size, load/store queues size, etc. The devised predictor is easily integrated in most system schedulers, providing the ability to periodically determine whether a certain thread is running in the most efficient core-architecture. The obtained experimental results show that the devised model is able to identify the correct core-architecture in a large majority of the cases, leading to average performance differences as low as 7% when compared with an oracle scheduling solution.

Keywords: Performance estimation · Linear regression model · Heterogeneous systems · Single-ISA architecture · Many-core processor · Application scheduling

1 Introduction

Advances in processor design have recently pushed for the development of heterogeneous processors, in order to tackle the power and memory walls. In particular, by relying on appropriate and different core architectures, it is possible to efficiently leverage Memory-Level Parallelism (MLP) and Instruction-Level Parallelism (ILP) [7,8] such as to minimize power and energy consumption with a reduced performance loss. However, exploiting heterogeneity often requires the development of efficient scheduling mechanisms, in order to anticipate the performance gains due to the migration of an application from one core to another, or to the morphing of a given core, which can be achieved by means of clock/power gating or by relying on reconfigurable technologies.

In particular, driven by the introduction of the ARM big.LITTLE heterogeneous processor [1] (although not exclusively), intensive research has recently been put forth in the exploitation of heterogeneous processor systems composed

This work was partially supported by national funds through Fundação para a Ciência e a Tecnologia (FCT), under project UID/CEC/50021/2013.

© Springer International Publishing AG 2017
I. Dutra et al. (Eds.): VECPAR 2016, LNCS 10150, pp. 101–111, 2017.
DOI: 10.1007/978-3-319-61982-8_11

of multiple in-order and out-of-order cores, by developing methodologies to manage the allocation of tasks to cores. For example, Patsilaras et al. [9] described a Chip Multi-Processor (CMP) with two core architectures, where one is tuned for exploiting MLP and the other for ILP. To manage the application allocation, the authors make use of on-line sampling techniques performed on both core architectures, as well as a heuristic algorithm based on the detection of clustered Last-Level Cache (LLC) misses.

A similar methodology was employed by Kumar et al. [8], although relying on a two-stage approach. During the first stage (*sampling*), applications are permuted over all core architectures in order to obtain a set of per-core statistics, retrieved from Hardware Counter (HCs). In a second stage, the gathered statistics are used to predict which core is the best suited for each application. Although the authors consider the possibility of using more than two different core types, they still require periodic on-line sampling of all application-core permutations, which a slow process and requires the system to operate sub-optimally during such periods.

Naturally, various attempts have been made to avoid this slow sampling process. For example, Shelepov et al. [13] described a computational system where the scheduler is supplied with application signatures, obtained through off-line analysis. However, this requires all applications to be re-compiled specifically for such a system, which is not always feasible. Saez et al. [11] use the count of LLC misses to grossly estimate the speedup factor without having to sample all application-core permutations. However, other parameters (e.g., core width), which cause many different interactions affecting application performance, cannot be properly described by just analyzing cache miss rates. Craeynest et al. [15] took a similar approach, by deriving an HC based simplified model to estimate performance differences between *small* in-order and *big* out-of-order cores. Based on this model, the authors developed a system scheduler to regularly estimate the performance of running applications on the alternate core and decide whether a core switch is worthwhile. However, this approach is constrained to two core types and can only take into account a small subset of architectural changes, namely in Re-order Buffer (ROB) size and issue width. Hence, other parameters (e.g. the cache hierarchy) are not correctly predicted. Taking this into consideration, Pricopi et al. [10] developed an ARM big.LITTLE specific prediction model that is able to take into account more architectural parameters, using a mixture of HC statistics and offline analysis. However, in addition to the requirement of an offline analysis, it still only considers two possible core variations at once.

Other scheduling approaches have also been proposed based on the similarity between the considered application and a previously known group or class of applications. For example, Delimitrou and Kozyrakis [4] proposed the use of a collaborative filtering technique on large data centers to schedule the application, by identifying similarities with previously known applications. However, such method requires an offline sampling process across a large set of server configurations. Other approaches have also been proposed for the specific purpose of guaranteeing the quality of service on ARM big.LITTLE systems (e.g., [5,6,16]).

However, such strategies focus on applications with real-time constraints and cannot be applied to the general case.

In accordance, this paper addresses the identified issues and limitations by proposing a new low-overhead and architecture-independent method to derive adaptable performance models. The devised models estimate the attainable performance over a large range of varying micro-architectural parameters and can be used both at a hardware-level or as a software module integrated into the OS scheduler. The considered approach makes use of a Linear Regression Model based on several commonly available HCs. In order to fully illustrate the proposed method, an example model based on out-of-order cores with different cache hierarchies, Re-order Buffer (ROB), Load Queue (LQ) and Store Queue (SQ) sizes was derived. The resulting model was then cross-validated with a set of 81 different core types using the PARSEC benchmark suite [2] and the micro-architectural simulator Sniper [3]. The proposed model is shown to be highly accurate and, when integrated with a system scheduler, is able to obtain performance errors below 2.2% and 6.8% for an heterogeneous processor featuring 2 and 11 different cores, respectively.

The remaining of this manuscript is organized as follows. Section 2 presents the proposed performance modeling approach, which relies on typically available processor HCs to construct a linear regression model. By considering a set of important architectural parameters that influence both memory- and instruction-level parallelism, a logarithmic-linked function is derived to estimate the average performance of an application (measured using the Cycles Per Instruction (CPI) metric) when the execution is migrated from one core to another. Section 3 presents the experimental results, by considering the difference between the real and the predicted performances of a set of applications extracted from the PARSEC benchmark suite. Finally, Sect. 4 concludes the paper by highlighting the main contributions.

2 Performance Modeling

Most current processors are equipped with multiple HCs that can be configured to measure various runtime statistics (e.g., cycle counts, retired instructions, cache misses), which can then be used to infer the application performance [11,13,15]. Such information allows for the development of intelligent software and/or hardware modules, capable of scheduling running applications to the most appropriate core architectures and/or adapting the characteristics of each core according to the scheduled application's computational requirements. Hence, it is herein considered that, during program execution, a set of HCs are measured at a *source core*, in order to characterize the current application phase. Based on such information, the devised system is able to predict the attainable performance on a *target core*, in order to support a decision on whether to move the thread to a different core or to apply any core morphing techniques. Like previous cross-core performance models, the proposed methodology assumes that any cross-thread interaction effects (i.e., cache sharing or synchronization) are core-independent, such that they manifest on all target cores similarly to the source core, reducing the modeling difficulty considerably.

Hence, this manuscript leverages the correlation between HC statistics and application performance in order to derive cross-core performance models. To attain such a goal, a Linear Regression Model (LRM) is adopted, which allows accurate performance predictions across hypothetical changes on several micro-architectural parameters, given an initial representative training set. Moreover, considering that the *retired instruction count* is an easy-to-measure and core-independent runtime statistic, it is used to normalize all runtime statistics into an application-independent scale that is easier to work with. As a result, CPI becomes an obvious choice for performance metric and is therefore used as the LRM dependent variable, since it can also be easily measured and is already normalized by the instruction count.

Furthermore, in order to improve the quality of the model, a logarithm link function is used. This is a natural approach, not only because the CPI metric is always positive, but also because experimental evaluation has shown that the original model's residual distribution is log-normal. Accordingly, since normally-distributed residuals are preferable in order to ensure that the least-squares estimator matches the maximum-likelihood estimator (as the latter has better statistical properties [12]), the proposed model is constructed in order to estimate the logarithm of the CPI at a target core (tgt), $\log(\hat{CPI}_{tgt})$, by relying on the perceived performance at a source core (src) and on a set of HCs that are highly correlated with the architectural differences between source and target cores. Hence, when applying the adopted LRM, the following performance estimation equation is obtained:

$$\log(\hat{CPI}_{tgt}) = \beta_0 + \beta_1 \log(CPI_{src}) + \sum_{i=1}^{N} \beta_{i+1} x_i \,, \tag{1}$$

where $\hat{CPI}_{tgt}$ represents the estimated CPI at the target core, β_i are model coefficients (in particular, β_0 represents the constant or intercept term), $\log(CPI_{src})$ represents the logarithm of the CPI measured in the source core, and $x_1, \cdots, x_N$ represent the set of N regression terms obtained by coupling the statistics gathered by using HCs with the micro-architectural parameter variations. Each regression term x_i is herein considered to express the product of the variation Δp of a given micro-architectural parameter p between the source (p_{src}) and target (p_{tgt}) cores $(\Delta p = p_{tgt} - p_{src})$, with a runtime statistic S_i, normalized by the retired instruction count I:

$$x_i = \frac{S_i}{I} \Delta p_i \,. \tag{2}$$

Concerning the selection of regression terms, it is important to note that, although the model accuracy generally increases with the introduction of more regression terms, this leads to an increase in model complexity and possibly to over-fitting, reducing its effectiveness when applied to new (i.e., unobserved) applications. It is therefore important to carefully select the minimum number

Table 1. Description of the considered set of core parameters, together with their dominant effects concerning the attained performance.

Architecture parameter	Description	Dominant effects
$L\{1,2,3\}size$	Total size of caches L1, L2 and L3	Impacts the cache hit rate, significantly impacting the memory access latency
$LQsize$	Load Queue size	When full, generates structural hazards for new load instructions, causing pipeline stalls at the issue stage
$SQsize$	Store Queue size	When full, generates structural hazards for new store instructions, causing pipeline stalls at the issue stage
ROB	Re-order Buffer size	When full, generates structural hazards, leading to stalls at instruction issue
W	Core issue, dispatch and commit Width	Affects the peak instruction throughput at issue, dispatch and commit stages

of terms that allow attaining an effective modeling of the dominant effects of all architectural parameters of interest. This procedure can be automated using statistical methods for automatic regressor choice (e.g., Lasso [14] or Elastic Net [17]), which provide the means for an automatic search over the regressor space in order to retrieve the most adequate architectural parameters and runtime statistics. Nevertheless, because the number of architectural parameters herein considered is not too large such approaches are not strictly necessary.

In order to obtain a generic model that covers a representative set of parameters, and simultaneously shows the flexibility of using a LRM to predict performance differences between different cores, a highly heterogeneous many-core CMP is herein considered as an example proof of concept, including many different out-of-order architectures of varying cache sizes (although limited to equal sized L1 instruction and data caches), issue widths, ROB sizes, as well as different load and store queue sizes (modeled as two separate queues). The set of considered parameters and their dominant effects are summarized in Table 1. Accordingly, it is necessary to choose runtime application-dependent statistics that are most correlated with the dominant effects of each micro-architectural parameter being varied. In order to choose between different runtime statistics that explain similar effects, their impact on the model prediction quality was evaluated by relying on the Sniper Multi-Core Simulator [3] to provide accurate simulations of several x86 micro-architectures. To analyze the results, the t-statistic (i.e., significance) was used, as well as the coefficient of determination R^2 of the resulting model. Nonetheless, the ease of measuring the various possible statistics in real hardware was also taken into account. The result of this analysis is presented in Table 2. As can be seen, all the chosen statistics correlate with at least one of the dominant effects mentioned in Table 1.

Table 2. Runtime statistics subset (most-relevant) for each processor parameter, the corresponding effect, and the maximum observed absolute t-Statistic value. Boldfaced t-Statistic values represent the variables used in the final model.

Hardware counter	Correlates with	t-Stat.
▶ **Core Width** (W) related architectural parameters		
I: Instruction Count	Peak performance	**11.36**
$Hdep$: Data Hazards at dispatch	Instruction interdependency	**10.12**
▶ **ROB Size** (ROB) related architectural parameters		
$Hrob$: Hazards due to full ROB	ROB occupancy	**6.62**
$Hdep$: Data Hazards at dispatch	Instruction interdependency	**6.75**
▶ **Load Queue Size** ($LQsize$) related architectural parameters		
LD: Load Uops Count	Load queue usage rate	**26.81**
Hlq: Hazards due to full LQ	Load queue usage rate	18.11
▶ **Store Queue Size** ($SQsize$) related architectural parameters		
ST: Store Uops Count	Store queue usage rate	**19.51**
Hsq: Hazards due to full SQ	Store queue usage rate	16.71
▶ **Cache Sizes** ($L\{1,2,3\}size$) related architectural parameters		
$L\{1,2,3\}miss$: Cache miss Count	Memory access latency	**7.80**
LD: Load Uops Count	Cache access rate	3.81
ST: Store Uops Count	Cache access rate	4.12

To better illustrate the considered statistics, the maximum t-statistic value for a corresponding 3-coefficient model ($N = 1$) is also presented, measured under the same experimental methodology as the results that will be presented in Sect. 3. For comparison purposes, some statistics that were left out from the proposed model are also shown. As can be seen, their corresponding t-statistic values are considerably lower than that of the selected HC based statistics (presented in boldface).

To further evaluate the relationship between the architecture parameters and the identified statistics, each of the considered architectures parameters were varied (one at a time), and their impact on each of the considered statistics was measured. The subsequent analysis was conducted by means of a set of scatter plots containing the parameters variation (x-axis) and the variables of interest (y-axis). It was then observed that some of the variables present a non-linear correlation with the corresponding architecture parameter. To model such cases, a Taylor series expansion was used. Hence, the following simple, but still highly representative, 14-term LRM was obtained:

$$
\begin{aligned}
\log(\hat{CPI}_{tgt}) = \beta_0 &+ \beta_1 \log(CPI_{src}) + \beta_2 \ L1miss_n \ \Delta L1size + \\
&\beta_3 \ L2miss_n \ \Delta L2size + \beta_4 \ L3miss_n \ \Delta L3size + \\
&LD_n \ (\beta_5 \ \Delta LQsize + \beta_6 \ \Delta LQsize^2) + \\
&ST_n \ (\beta_7 \ \Delta SQsize + \beta_8 \ \Delta SQsize^2) + \\
&Hdep_n(\beta_9 \ \Delta ROB + \beta_{10} \ \Delta W) + \\
&\beta_{11} \ Hrob_n \ \Delta ROB + \beta_{12} \ \Delta W + \beta_{13} \ \Delta W^2 \ .
\end{aligned}
\tag{3}
$$

Upon obtaining the above defined LRM, the β_i coefficients were estimated by training the LRM with observations obtained by running a representative set of benchmarks on all core variations of interest, resulting in a linear number of models (i.e., one per source core). It should be noticed that the number of terms in (3) was chosen such as to allow an overall minimization of the estimation error when applied to an independent group of benchmarks (i.e., different from the dataset used to estimate the model parameters). Moreover, when considering two models with similar error values, the one with the lowest number of terms was chosen.

3   Experimental Results

In order to properly evaluate the developed cross-core performance model, the Sniper Multi-Core Simulator [3] was used, to provide accurate simulations of several x86 micro-architectures. Hence, a vast set of core variations was described in this simulation framework, by varying several highly important micro-architecture and cache organization parameters, as depicted in Tables 3 and 4. In accordance, a total of 81 different core variations were simulated, allowing an effective modeling of the interaction between the considered parameters.

To ensure the representativeness of the devised model when considering multiple types of workloads, the PARSEC [2] benchmark suite was chosen for its training and validation procedures. For such purpose, simulator-specific *magic* instructions were added to each of the eleven PARSEC benchmarks, in order to define the appropriate simulation Region of Interest (ROI) for each benchmark, therefore excluding the initialization and shutdown phases, since these depend almost solely on the systems outside of the processor's control (e.g., hard drive data access latency and bandwidth) and are therefore uninteresting from an architectural point-of-view.

The benchmarks were then executed to completion using the predefined "small" input set on each of the 81 different processors, with the pre- and post-ROI sections simulated in fast-forward mode in order to reduce the processing time. All runtime statistics required by the model were measured during the execution and stored for later processing.

Table 3. Considered cache hierarchy variations (associativity, set count, and total size in KB); The block size was set fixed and equal to 64 Bytes.

Cache level	Configuration	Associativity	Set count	Total size
▶ L1-D ▶ L1-I	Small	2	8	1 KB
	Medium	2	16	2 KB
	Large	4	32	8 KB
▶ L2	Small	4	32	8 KB
	Medium	8	64	32 KB
	Large	8	256	128 KB
▶ L3	Small	8	1024	512 KB
	Medium	16	2048	2048 KB
	Large	16	8192	8192 KB

Table 4. Considered architecture variations

Architecture parameter	Considered values
▶ Load Queue size ($LQsize$)	1; 5; 10
▶ Store Queue size ($SQsize$)	1; 5; 10
▶ Re-order Buffer size (ROB)	32; 64; 128
▶ Core issue, dispatch and commit Width (W)	1; 4; 8

3.1   Model Validation

Since the model assumes the representativeness of the training set for all possible applications and cores, it makes sense to use as much information as possible during its validation. Therefore, a leave-one-out cross-validation approach was adopted, such that one random application was removed from the training set in each iteration, and subsequently used for model validation. Moreover, to guarantee correctness in the evaluation procedure, none of the applications used in the training procedure were used for the validation procedure. Finally, in order to further illustrate the quality of the model, multiple goodness-of-fit measures were calculated for each of the 81 individual source core models.

Figure 1 presents the CPI normalized prediction over all considered architecture variations, represented as a Tukey box-plot for each benchmark. As can be observed, the model provides accurate predictions over a wide range of application characteristics for all considered core parameters. On the other hand, it can also be observed that the largest prediction error occurs for the *canneal* and *streamcluster* applications, which is explained by the fact that these benchmarks comprehend a larger inter-phase variation of the observed CPI. Such a variation could be explained (in future work) by evaluating the error across application phases, instead of evaluating across the whole application execution.

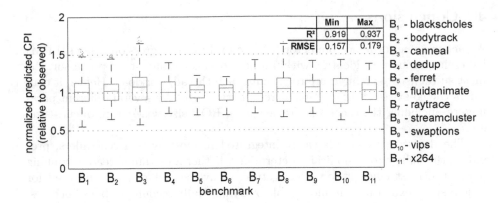

Fig. 1. Predicted CPI (with cross-validation) for all considered architecture variations, with the minimum and maximum values of the coefficient of determination (R^2) and of the Root Mean Square Error ($RMSE$) obtained for all models.

Table 5. Scheduler validation test results

# of Cores N	2	3	6	11
Random Scheduler CPI	1.67	1.67	1.67	1.67
Best/Oracle Scheduler CPI	1.38	1.22	1.07	1.03
Proposed Model Scheduler CPI	1.41	1.28	1.15	1.10
Relative Error (Proposed vs. Oracle)	2.17%	4.92%	7.48%	6.80%

An F-test of overall significance [12] was also performed on all models, in order to evaluate whether a simple intercept-only fit would be statistically indistinguishable from the proposed models. The obtained results showed a p-Value of 0 for all cases, which fulfills this basic quality requirement.

Lastly, a scheduler-specific validation test was performed, which evaluates whether the proposed model could effectively predict the most efficient core for each application. Hence, for each iteration of the test, a permutation of one source core and $N - 1$ alternative target cores was picked at random. The model was then used to predict the best core (minimum CPI) for each application, out of the N possible choices. The observed CPI in the chosen core was then compared with the observed CPI of a scheduler using either a *random* or an *oracle* policy. A total of 891 000 iterations of this validation mechanism were executed using different values of N. The results, presented in Table 5, show that the model manages to estimate the correct core in a large majority of the cases. Furthermore, when the proposed model performs an incorrect guess, only a reduced performance loss is observed when compared to the *oracle* case.

4   Conclusions

An accurate performance predictor based on a Linear Regression Model (LRM) is herein proposed to identify, within a heterogeneous many-core processor, the most suitable core-architecture to execute each thread/workload. Hence, it considers the co-existence of multiple cores, characterized by several different parameters, including the cache size, issue-width, ROB size, load/store queues size, etc.

The devised predictor is easily integrated in most system schedulers, providing the ability to periodically determine whether a certain thread is running under the most efficient core-architecture. Conversely, it can also be used for design space exploration in morphable or dynamically reconfigurable structures, not only to determine when the processing architecture should be reconfigured, but also to determine the corresponding set of parameters.

The experimental evaluation showed that the devised model is able to identify the correct core-architecture in a large majority of the cases, leading to average performance differences as low as 7% when compared with the *oracle* solution.

The offered flexibility makes the devised model easily adaptable to other optimization metrics besides the considered CPI. As an example, an energy estimation model can be easily implemented, in order to obtain energy/power-aware scheduling schemes.

References

1. big.LITTLE Technology: The Future of Mobile. Technical report, ARM (2011). https://www.arm.com/files/pdf/big_LITTLE_Technology_the_Futue_of_Mobile.pdf
2. Bienia, C.: Benchmarking Modern Multiprocessors. Ph.D. thesis, Princeton University, Princeton, NJ, USA (2011)
3. Carlson, T.E., Heirman, W., Eyerman, S., Hur, I., Eeckhout, L.: An evaluation of high-level mechanistic core models. ACM Trans. Archit. Code Optim. (TACO) **11**(3), 28:1–28:25 (2014)
4. Delimitrou, C., Kozyrakis, C.: Paragon: Qos-aware scheduling for heterogeneous datacenters. In: Proceedings of the Eighteenth International Conference on Architectural Support for Programming Languages and Operating Systems, ASPLOS 2013, pp. 77–88. ACM, New York (2013)
5. Gaspar, F., Taniça, L., Tomás, P., Ilic, A., Sousa, L.: A framework for application-guided task management on heterogeneous embedded systems. ACM Trans. Archit. Code Optim. **12**(4), 42:1–42:25 (2015)
6. Imes, C., Kim, D.H., Maggio, M., Hoffmann, H.: POET: a portable approach to minimizing energy under soft real-time constraints. In: Proceedings of the Real-Time and Embedded Technology and Applications Symposium (RTAS), pp. 75–86. IEEE (2015)
7. Kumar, R., Farkas, K.I., et al.: Single-ISA heterogeneous multi-core architectures: the potential for processor power reduction. In: 36th Annual IEEE/ACM International Symposium on Microarchitecture, MICRO 36, pp. 81–92. IEEE Computer Society (2003)

8. Kumar, R., Tullsen, D.M., et al.: Single-ISA heterogeneous multi-core architectures for multithreaded workload performance. SIGARCH Comput. Archit. News **32**(2), 64–75 (2004)
9. Patsilaras, G., Choudhary, N.K., Tuck, J.: Efficiently exploiting memory level parallelism on asymmetric coupled cores in the dark silicon era. ACM Trans. Architect. Code Optim. (TACO) **8**(4), 28:1–28:21 (2012)
10. Pricopi, M., Muthukaruppan, T.S., et al.: Power-performance modeling on asymmetric multi-cores. In: 2013 International Conference on Compilers, Architecture and Synthesis for Embedded Systems (CASES), pp. 1–10 (2013)
11. Saez, J.C., Prieto, M., et al.: A comprehensive scheduler for asymmetric multicore systems. In: 5th European Conference on Computer Systems, EuroSys 2010, pp. 139–152. ACM (2010)
12. Seber, G.A.F., Lee, A.J.: Linear Regression Analysis. Wiley, New York (2003)
13. Shelepov, D., Saez Alcaide, J.C., Jeffery, S., Fedorova, A., Perez, N., Huang, Z.F., Blagodurov, S., Kumar, V.: HASS: a scheduler for heterogeneous multicore systems. SIGOPS Oper. Syst. Rev. **43**(2), 66–75 (2009)
14. Tibshirani, R.: Regression shrinkage and selection via the lasso. J. Roy. Stat. Soc. Ser. B (Methodological) **58**, 267–288 (1996)
15. Van Craeynest, K., Jaleel, A., et al.: Scheduling heterogeneous multi-cores through performance impact estimation (PIE). In: 39th International Symposium on Computer Architecture, ISCA 2012, pp. 213–224. IEEE Computer Society (2012)
16. Zhu, Y., Halpern, M., Reddi, V.J.: Event-based scheduling for energy-efficient QoS (eQoS) in mobile web applications. In: Proceedings of the International Symposium on High Performance Computer Architecture (HPCA), pp. 137–149. IEEE (2015)
17. Zou, H., Hastie, T.: Regularization and variable selection via the elastic net. J. Roy. Stat. Soc. Ser. B (Statistical Methodology) **67**(2), 301–320 (2005)

On the Acceleration of Graph500: Characterizing PCIe Overheads with Multi-GPUs

Mayank Daga$^{(\boxtimes)}$

AMD Research, Advanced Micro Devices, Inc., Sunnyvale, USA
Mayank.Daga@amd.com

Abstract. Graphics Processing Units (GPUs) have fundamentally altered the approach to parallel computing despite the substantial PCIe overheads that they manifest. In order to maximize performance-per-dollar, systems are now being deployed with multiple GPUs in the same node. However, multiple GPUs exacerbate the PCIe overheads by inflicting additional data-movement performance penalties when moving non-local data.

In this paper, we first evaluate the PCIe performance loss that occurs due to improper affinity between CPUs and GPUs, using a PCIeBandwidth benchmark specifically developed for systems with multiple GPUs. Our experiments demonstrate that the performance loss can be up to 2.5× on a single GPU and up to 4.4× when four GPUs are used. We then leverage our learnings from the PCIe studies to optimize and accelerate the Graph500 benchmark on a 4-GPU, multi-socket system. Our optimization techniques include binding the CPU threads to appropriate cores as well as the careful partitioning of data for every GPU. We achieve a speedup of 1.8× over a single GPU implementation.

1 Introduction

The exigent demands of emerging applications to maximize performance while staying under power and thermal constraints have made graphics processing systems (GPUs) ubiquitous [8,9]. Since GPUs have traditionally resided on PCI Express (PCIe), additional overheads are incurred for host-to-GPU data transfers and vice versa. As a consequence, GPU applications are oftentimes bottlenecked by the PCIe data transfers [6]. Despite this fact, GPUs have achieved immense popularity due to a unique combination of performance and energy efficiency. GPUs have also been recognized to play an important role on the path to extreme scale computing as evident by the fact that half of the top ten supercomputers on the Top500 list use GPUs as accelerators [1].

In order to maximize performance-per-dollar, systems are now being deployed with multiple GPUs. However, multiple GPUs bring in additional challenges particularly with respect to optimal PCIe performance. This is because of the complex mapping between the GPUs which require data across the PCIe and the CPU cores which are responsible for doing the direct memory access (DMA) of data. In such systems, data-transfers occur at full PCIe bandwidth between *local* CPU

© Springer International Publishing AG 2017
I. Dutra et al. (Eds.): VECPAR 2016, LNCS 10150, pp. 112–120, 2017.
DOI: 10.1007/978-3-319-61982-8_12

cores and GPUs. However, data-transfer to a remote GPU is subject to a significantly worse bandwidth because of additional on-chip interconnects. The onus of placing data on appropriate DMA nodes lies on the application developer.

In this paper, we first evaluate the cost of moving data from the host to GPU at various combinations of mapping between the CPU cores and GPUs. For doing so, we use an indigenous `PCIeBandwidth` benchmark which allows us to (i) bind data to a particular DMA node, (ii) bind a CPU thread to a particular core and then (iii) use that thread to transfer data to any particular GPU or multiple GPUs. We then leverage our learnings from the PCIe studies to optimize the Graph500 benchmark [2]. Graph500 uses breadth-first search (BFS) as its main kernel and tracks the fastest data-intensive supercomputers in the world. Almost half of the total execution time in Graph500 is spent in moving data to the GPU [5,10]. Therefore, managing the data- and thread-bindings is imperative to achieve good performance. Our implementation of Graph500 uses the `hybrid++` algorithm which partitions the computation between CPU and GPU and hence, good PCIe performance is crucial [7,11]. We implement the following optimization techniques: (i) partition data in chunks for each GPU, (ii) manually map those chunks to local DMA nodes and, (iii) bind CPU threads to a particular core which is local to the GPU.

The PCIeBandwidth benchmark demonstrates that the performance degradation due to incorrect mapping between CPU cores and GPU can be up to 2.5× when a single GPU is used and up to 4.4× when four GPUs are used. Our optimized Graph500 implementation on a 4-GPU multi-socket system achieves a speedup of 1.8× compared to a single GPU implementation.

The rest of the paper is arranged as follows. Section 2 provides a background on the Graph500 benchmark and the algorithm that we use to compute BFS. Section 3 describes our experimental setup followed by the characterization studies of PCIe in Sect. 4. Section 5 presents the optimizations and evaluation of the multi-GPU Graph500 implementation. Section 6 presents the conclusion of this work.

2   The Graph500 Benchmark

Graph500 uses a breadth-first search (BFS) kernel to rank the top data-intensive supercomputers in the world. The benchmark provides freedom to the developers to use any custom BFS algorithm for the purposes to computing the score in the form of giga-transferred-edges-per-second (GTEPS). We use the hybrid++ BFS algorithm in our Graph500 implementation [4,7].

The hybrid++ algorithm is suitable for heterogenous processors as it enables us to choose between a combination of traversal directions (top-down or bottom-up) and the platform of execution (CPU or GPU). The top-down traversal is a serial algorithm that executes best on CPUs. Whereas the bottom-up traversal exposes immense parallelism and is suitable for GPUs. The hybrid++ algorithm uses an online heuristic to seamlessly choose the appropriate algorithm and the suitable core for every iteration of BFS. The heuristic leverages input graph characteristics as well as traversal information from prior BFS iterations to make optimal decisions. A high-level illustration of hybrid++ is shown in Fig. 1.

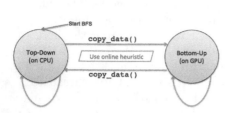

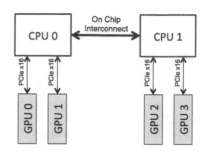

Fig. 1. High-level block diagram illustrating hybrid++ BFS algorithm [7].

Fig. 2. High-level block diagram illustrating a dual socket multi-GPU system with 4 GPUs connected across PCIe x16 Gen3.

Since, hybrid++ requires a data-copy whenever there is a change in the BFS traversal algorithm, achieving good PCIe performance is imperative for an efficient overall execution of Graph500. Therefore, understanding and characterizing the PCIe effects with multiple GPUs is key to the acceleration of Graph500.

3 Experimental Setup

The schematic diagram of the system we use for our experiments is shown in Fig. 2. The system consists of a Intel® Xeon® E5-2667 v3 CPU which has two sockets. There are four PCIe Gen3 x16 lanes with two such lanes connected to each socket. The GPUs reside on the PCIe with one GPU per PCIe. Therefore, as per the figure GPUs 0 and 1 are local to CPU 0 and GPUs 2 and 3 are local to CPU 1. Moving data to/from a remote GPU occurs over the on chip interconnect and hence, adversely affects performance.

We performed all our experiments on an AMD FirePro™ S9150 GPU with ECC disabled. Figure 6 details its important characteristics. The host machine uses 64 GB of DDR3-2133 SDRAM. The GPU was programmed using OpenCL™ v2.0 with the AMD APP SDK v3.0 and AMD FirePro driver v15.20. The operating system was a 64-bit version of CentOS 6.4, kernel version 2.6.32–358.23.2.

The input for Graph500 is a synthetic `rmat` graph [3]. We vary the number of nodes in the graph from 1- to 16-million with an edge-degree of 16.

For measuring the PCIeBandwidth, we use a 512 MB buffer of `float` data-type and the data is moved from host to the GPU. We used 256 threads per workgroup, and all of the performance numbers are an average of 1000 runs.

4 PCIe Bandwidth with Multiple GPUs

We developed a PCIe Bandwidth benchmark to understand the effects of mapping and affinity between the GPUs which need the data and the CPU cores

```
1   function AllocateAndRun() {
2       // create one thread per gpu
3       for g ∈ num_gpus do
4           std::thread new_thread (ThreadAllocateAndRun, /* function arguments */ );
5       end for
6   }
7   function ThreadAllocateAndRun( /* function arguments */ ) {
8       // bind this thread to a CPU core
9       pthread_setaffinity_np(pthread_self(), /* core to bind */ );
10      // allocate host buffer
11      TYPE *hostMem = new TYPE[SIZE];
12      // allocate device buffer
13      cl_mem devMem = clCreateBuffer( /* function arguments */ );
14      // move data across PCIe and measure the bandwidth
15      clEnqueueWriteBuffer(..., devMem, hostMem, ...);
16  }
```

Fig. 3. Pseudocode for PCIe Bandwidth benchmark.

which DMA that data. The pseudocode for the benchmark is shown in Fig. 3. Using PCIe Bandwidth we can control the following: (i) mapping data to a particular DMA node, (ii) binding CPU threads to particular cores, (iii) which GPU in the system to use, and (iv) the number of GPUs to use.

We compute the PCIe bandwidth achieved using various mappings of CPU cores and GPUs and characterize the effects of moving local and non-local data across the PCIe in a multi-GPU system. For example as per Fig. 2, if data is mapped to CPU 0 and moved to GPU 0 then the transfer is local but if it is moved to GPU 2 then the transfer is remote and occurs via the on-chip interconnect.

Figure 4 demonstrates the PCIe bandwidth achieved with both local and remote data-transfers when one, two or four GPUs are used. Using a single GPU and moving data to a local GPU, we achieve a unidirectional bandwidth of 12.3 GB/s. We normalize our results to this number, which is the best-case, and present them in Fig. 4a. From the figure, we note that performance is consistent no matter which GPU among the four GPUs in the system are used. The difference between local and remote data-transfers is 2.5×. This is because a remote data-transfer adds the latency of on-chip interconnect.

Figure 4b illustrates the PCIe bandwidth achieved when using two GPUs residing on two different nodes, e.g., one among GPUs 0 or 1 and one among GPUs 2 or 3. To achieve the best bandwidth with two GPUs, both the GPUs need to do local data-transfers. This means that the data has to be partitioned and allocated half on each memory node. From the figure, we note that the local bandwidth achieved with two GPUs is 91% of the bandwidth achieved with one GPU due to inherent system overheads. When either of the two GPUs is doing a remote transfer, its bandwidth reduces significantly just as in the case of a single GPU. When both the GPUs are doing remote transfers, the bandwidth achieved is 2.6× lower than the maximum possible.

Figure 4c illustrates the PCIe bandwidth achieved when using two GPUs residing on the same node, e.g., either GPUs 0 and 1 or GPUs 2 and 3. Best bandwidth is achieved when both the GPUs are doing local transfers. However, the bandwidth achieved by both the GPUs is not equal; GPU 3 achieves 6% lower bandwidth than

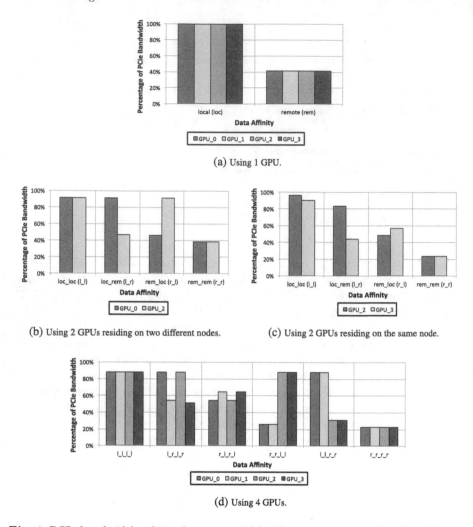

Fig. 4. PCIe bandwidth achieved as measured by the `PCIeBandwidth` benchmark using various combinations of 4 GPUs. All the results are normalized to the bandwidth achieved by a single GPU when moving data local to its node. Local (or `loc (l)`) means the GPU closer to data is used. Remote (or `rem (r)`) means that the GPU farther away from the data is used.

GPU 2. Bandwidth achieved when either of the two GPUs is remote is also erratic. From the figure, when GPU 3 is remote, bandwidths achieved by GPUs 2 and 3 are 81% and 41% of peak, respectively. However, when GPU 2 is remote the bandwidths achieved are only 48% and 57% of peak, respectively. The reason for this is the contention of resources on the same memory node while carrying out the DMA to GPUs. When both the GPUs are remote, the bandwidth achieved is 2.6× lower than than achieved by a single GPU. Therefore, if two GPUs are required to be used, the application developer should ensure that both the GPUs reside on different sockets in a multi-socket system.

Figure 4d illustrates the PCIe bandwidth achieved when using all four GPUs in the system. As in other cases, the best bandwidth is achieved when all the GPUs are accessing local data. The bandwidth achieved by a single GPU when all the GPUs are active is 88% of what is achievable when only one GPU is active. From the figure, we note that inconsistent bandwidths are achieved at different combinations of local and remote data transfers due to the underpinnings of the system which are hidden from the application programmer. When all the GPUs are remote, the bandwidth achieved is worst and is 4.4× lower than that achieved by a single GPU.

From the above results, it is clear that manually managing the data and thread bindings is vital to extract efficient performance when using multiple GPUs. Not controlling the data binding allows the runtime and operating system to freely modify the bindings without programmer knowledge, thereby resulting in suboptimal performance, as shown in Fig. 4. A particular feedback to the runtime developers is to make the process of binding threads and data easier by providing APIs to do so.

5   Graph500: Optimization and Evaluation

In this section we describe our optimization strategies for the Graph500 benchmark. Graph500 requires the BFS tree that is generated as part of the search to be preserved as final output. The buffer containing the BFS tree is copied to and from the host in order to keep an updated copy of the resulting output.

All the GPUs computing BFS access the BFS tree. However, due to the data-parallel nature of bottom-up BFS they access different regions of the buffer thereby, allowing the buffer to be partitioned among the GPUs. As we note in Sect. 4, for efficient PCIe performance each chunk of the buffer should be copied to the local GPU. Therefore, we first create as many chunks of the BFS tree buffer as the number of GPUs and then map each chunk to the closest CPU node which will do the DMA. For mapping the chunks on the host, we use Pthreads to create a new host thread for every GPU and set its affinity to the core closest to that GPU. For example, a thread t_0 is created for GPU 0 and its affinity is set to core_0. Similarly, for GPU 2, a thread t_2 is created and its affinity is set to core_8 because each CPU has 8 cores in our system. Hence, t_2 is bound to CPU 1. Once the affinities are set, the same threads are used to allocate the GPU buffers using OpenCL APIs and then DMA the data to their local GPU. Since the threads are manually bound to CPU cores, they use the DMA engines on the same node as the CPU thereby, ensuring that the data is moved to the local GPU. Figure 5 illustrates this optimization process.

In Fig. 7, we plot the time taken to move the data to and from the GPU, i.e., the copy time, and the time taken to do the actual search on the GPU, while varying the number of GPUs. The GPU time is measured using OpenCL event APIs and copy time is measured using clock_gettime() on the host. From the

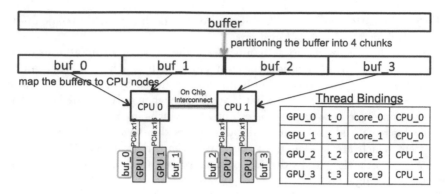

Fig. 5. Partitioning and mapping of the BFS tree buffer in Graph500 to achieve efficient PCIe performance. The buffer is divided into 4 chunks because we are using 4 GPUs in the system.

CPU	Intel® Xeon® E5-2667v3
Cores	16 (8 on each socket)
GPU	AMD Firepro™ S9150
Compute Units (CU)	44
Core Clock Rate	930 MHz
GDDR5 Memory Clock Rate	1250 MHz
Memory Size	16 GB
Peak Memory Bandwidth	320 GB/s

Fig. 6. Overview of the test platform.

figure, we note that the copy time reduces by 3× when four GPUs are used. This is because as we increase the number of GPUs, the amount of data required to be moved becomes smaller due to partitioning of data, as shown in Fig. 5. The scaling of copy time is not perfectly linear because of the inherent runtime and operating system overheads, as outlined in Sect. 4. Similarly, the GPU time is reduced by 3.5× when using four GPUs thereby, demonstrating almost linear scaling. Therefore, our optimizations are quite effective in improving the performance of multi-GPU implementation of Graph500.

In Fig. 8, we demonstrate the impact of our optimizations as we increase the nodes and edges of the input graph, while varying the number of GPUs. For a single GPU, both baseline and optimized numbers are the same because all of our optimizations are targeted towards multiple GPUs. From the figure, we note that the optimizations always improve the performance. Speedup achieved by the optimizations alone can be up to 1.9× as shown for the 4 GPU run of the 1M-node graph. Overall, the speedup achieved compared to a single-GPU implementation is 1.8×, on average.

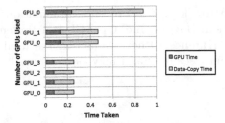

Fig. 7. Scaling of time spent on the GPU and to move data with increasing number of GPUs. This data is computed using the **rmat** graph with 8 M nodes and 96 M edges.

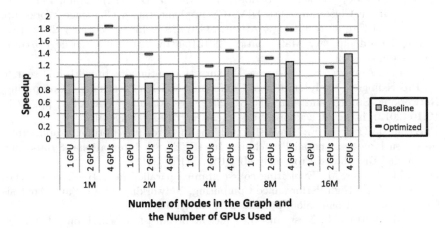

Fig. 8. Effect of optimizations on Graph500. Graph with 16M nodes could not be executed on one GPU due to memory limitations. Baseline is assumed to be performance of a single GPU but for the 16 M node-graph, baseline is performance of 2 GPUs

6   Conclusions

GPUs have become immensely popular for accelerating applications despite the PCIe overheads between the CPU and GPU. Nowadays, systems are being deployed with multiple GPUs on a single to maximize performance-per-dollar. However, multiple GPUs magnify the performance penalties of PCIe due to the possibility of moving data to non-local resources.

In this paper, we develop a novel PCIeBandwidth benchmark to characterize the PCIe overheads in a multi-GPU system. We also demonstrate the mechanisms for the efficient use of multi-GPU systems. Our experiments portray that a performance loss of up to 4.4× can occur while moving data to four GPUs using incorrect data- and thread-bindings. We then optimize the Graph500 benchmark on a multi-GPU, multi-socket, NUMA system and achieve a speedup of 1.8× over a single GPU.

AMD, the AMD Arrow logo, FirePro and combinations thereof are trademarks of Advanced Micro Devices, Inc. OpenCL is a trademark of Apple, Inc. used by permission by Khronos. Other product names used in this publication are for identification purposes only and may be trademarks of their respective companies.

References

1. The Top500 Supercomputer Sites. http://www.top500.org
2. The Graph500 Benchmark (2012). http://www.graph500.org
3. Bader, D.A., Meyerhenke, H., Sanders, P., Wagner, D. (eds.): Graph Partitioning and Graph Clustering -10th DIMACS Implementation Challenge Workshop, Georgia Institute of Technology, Atlanta, GA, USA, 13–14 February 2012, Proceedings, Contemporary Mathematics (2013). http://dblp.uni-trier.de/db/conf/dimacs/dimacs2012.html
4. Beamer, S., Asanović, K., Patterson, D.: Direction-optimizing breadth-first search. In: Proceedings of the International Conference on High Performance Computing, Networking, Storage and Analysis, SC 2012, Los Alamitos, CA, USA, pp. 12:1–12:10 (2012). http://dl.acm.org/citation.cfm?id=2388996.2389013
5. Checconi, F., Petrini, F.: Traversing trillions of edges in real-time: graph exploration on large-scale parallel machines. In: IEEE 28th International Symposium on Parallel Distributed Processing (IPDPS). IEEE (2014)
6. Daga, M., Nutter, M.: Exploiting coarse-grained parallelism in B+ Tree searches on an APU. In: High Performance Computing, Networking, Storage and Analysis (SCC), 2012 SC Companion, pp. 240–247, November 2012
7. Daga, M., Nutter, M., Meswani, M.: Efficient breadth-first search on a heterogeneous processor. In: Proceedings of the 2014 IEEE International Conference on Big Data (Big Data), October 2014
8. Daga, M., Feng, W., Scogland, T.: Towards accelerating molecular modeling via multiscale approximation on a GPU. In: Proceedings of the 1st IEEE International Conference on Computational Advances in Bio and medical Sciences (2011)
9. Owens, J.D., Houston, M., Luebke, D., Green, S., Stone, J.E., Phillips, J.C.: GPU computing. Proc. IEEE **96**(5), 879–899 (2008). http://www.idav.ucdavis.edu/publications/print_pub?pub_id=936
10. Ueno, K., Suzumura, T.: Highly scalable graph search for the Graph500 benchmark. In: Proceedings of the 21st International Symposium on High-Performance Parallel and Distributed Computing, HPDC 2012, New York, NY, USA, pp. 149–160 (2012). http://doi.acm.org/10.1145/2287076.2287104
11. Yasui, Y., Fujisawa, K., Goto, K.: NUMA-optimized parallel breadth-first search on multicore single-node system. In: BigData Conference, pp. 394–402. IEEE (2013). http://dblp.uni-trier.de/db/conf/bigdataconf/bigdataconf2013.html#YasuiFG13

Evaluation of Runtime Cut-off Approaches for Parallel Programs

Alcides Fonseca[✉] and Bruno Cabral

University of Coimbra, Coimbra, Portugal
amaf@dei.uc.pt, bcabral@dei.uc.pt

Abstract. Parallel programs have the potential of executing several times faster than sequential programs. However, in order to achieve its potential, several aspects of the execution have to be parameterized, such as the number of threads, task granularity, etc. This work studies the task granularity of regular and irregular parallel programs on symmetrical multicore machines. Task granularity is how many parallel tasks are created to perform a certain computation. If the granularity is too coarse, there might not be enough parallelism to occupy all processors. But if granularity is too fine, a large percentage of the execution time may be spent context switching between tasks, and not performing useful work.

Task granularity can be controlled by limiting the creation of new tasks, executing the workload sequentially in the current task. This decision is performed by a cut-off algorithm, which defines a criterion to execute a task workload sequentially or asynchronously. The cut-off algorithm can have a performance impact of several orders of magnitude.

This work presents three new cut-off algorithms: MaxTasksInQueue, StackSize and MaxTasksSS. MaxTasksInQueue limits the size of the current thread queue, StackSize limits the number of stacks in recursive calls, and MaxTasksSS limits both the number of tasks and the number of stacks. These new algorithms can improve the performance of parallel programs.

Existing studies have analyzed only two cut-off approaches at a time, each with its own set of benchmarks and machines. In this work we present a comparison of a manual threshold approach to 5 state-of-the-art algorithms (MaxTasks, MaxLevel, Adaptive Tasks Cutoff, Load-Based and Surplus Queued Task Count) and 3 new approaches (MaxTasksIn-Queue, StackSize and MaxTasksSS). The evaluation was performed using 24 parallel programs, including divide-and-conquer and loop programs, on two different machines with 24 and 32 hardware threads, respectively.

Our analysis provided insight of how cut-off algorithms behave with different types of programs. We have also identified the best algorithms for combinations of balanced/unbalanced and loop/recursive programs.

Keywords: Runtime · Cut-off Mechanism · Granularity · Multicore

© Springer International Publishing AG 2017
I. Dutra et al. (Eds.): VECPAR 2016, LNCS 10150, pp. 121–134, 2017.
DOI: 10.1007/978-3-319-61982-8_13

1   Introduction

Nowadays, making parallel programs faster is a manual process that relies on a lengthy trial-and-error process in order to achieve the best parameters. This process also requires a parallel programming expertise and knowledge of the program domain. Factors like thread or task creation, memory allocation and cache usage are fundamental in obtaining the best performance out of a multicore machine.

Although parallel programs can be more complex, we will focus on two of the most common parallelism patterns: for-loops and recursive programs. Parallelization of for-loops has been the basis of several wide-adopted frameworks, such as OpenMP [1]. Recursive programs have been the focus of parallelization in other frameworks such as Cilk [2] and ForkJoin [3].

In this work we consider task-based parallelization. Work is divided into several tasks that can be executed in parallel, on top of a work-stealing runtime. This kind of runtime executes one thread per CPU core, or hyperthread when supported, and stores a queue of pending tasks per thread. When a thread has no tasks to pop from the queue, it steals from the end of another thread queue. Tasks are used because they have less scheduling overhead than threads, as they do not require a system call, and they can be used to balance the load in irregular unbalanced programs.

In both for-loop and recursive parallelism patterns, choosing the best granularity is an important issue. In our evaluation, the same program could execute within seconds or within days, depending on the granularity selected. The granularity is defined as how many parallel tasks are created to perform a certain workload. Tasks are a representation of blocks of code that can execute independently on different threads.

If tasks are too coarse, there might not be enough tasks to occupy the hardware threads, resulting in unused hardware resources that could have been used to improve performance. If parallelism is too fine-grained, too many tasks will be created, imposing an overhead in task scheduling and management that will increase the duration of the execution. Achieving a good balance for all kinds of programs on different machines is thus crucial to achieve a good performance.

In this paper, we will study the most relevant state-of-the-art approaches to control the granularity of tasks at runtime. Alongside these, three new approaches will also be analyzed and studied. The goal of the study is to understand how these algorithms perform on parallel programs with different natures and on different machines, in order to understand which one should be used and when.

The remaining of the paper is organized as follows: Sect. 2 introduces the topic of granularity control; Sect. 3 details several approaches for controlling the cut-off threshold for parallelization; Sect. 4 evaluates cut-off algorithms; and finally, Sect. 5 presents the final conclusions of this study.

2   Granularity Control

Parallelizing compilers try to match parallel tasks with the layout of the underlying hardware. The static scheduling divides a loop in N chunks, one for each processor [4]. However, not all programs have this regular and static parallelism. Some programs have a more dynamic behavior, and the number of tasks changes across time. For these programs, runtime-based approaches are needed.

One common approach is to use a work-stealing scheduler, with Lazy Task Creation [5](LTC) as a granularity control mechanism. Potential parallel tasks might be executed inlined, or added to the work queue as a new task, according to different cut-off algorithms. These cut-off algorithms will be described in Sect. 3.

Cut-off algorithms have a great impact on the performance of programs. An OpenMP evaluation [6] has compared two approaches (*MaxLevel* and *MaxTasks*, explained in detail in Sect. 3) and it found differences of up to 3x of speedup, but could not provide any guidance of how to choose an ideal cut-off. Later, a second study focused on the granularity and found that *ATC*, also detailed in Sect. 3, was better than the worst approach, but not always better than the best approach [7].

Two other studies, one also within OpenMP [8] and another comparing OpenMP to other approaches [9], have shown differences between the two cut-off mechanisms on unbalanced task graphs. Given the random nature of the benchmark programs used, there was no information obtained over which of the two (*MaxLevel* and *MaxTasks*) approaches was better.

3   Cut-Off Mechanisms

A cut-off mechanism is an algorithm that decides whether a task will spawn new tasks for parallel work, or execute sequentially. Using LTC, it is possible to introduce a condition that stops the parallel execution of the program.

Different criteria have been proposed for deciding between parallel and sequential execution:

- **LoadBased** - The task will execute sequentially when all threads have work to perform in their queues. If there is at least one empty queue, it will execute in parallel [7].
- **MaxLevel**, or Maximum task recursion level - Divide-and-conquer algorithms create tasks in a tree-shaped structure. In order to avoid the creation of too many tasks, the cut-off limit may be defined by the depth of the recursion [6], which can be calculated by the number of ancestors of the running task. If the task has more than a parameterized threshold of ancestors, it will execute sequentially. This approach is more suitable for balanced programs, where all subtrees have the same depth.
- **MaxTasks**, or Maximum number of tasks - Using this approach, tasks are created until the total number of active tasks in all worker queues reaches a parameterized threshold [6]. After that point, all new computations are inlined instead of spawning another tasks. When the number of active tasks is lower,

new tasks can be created in parallel until the threshold is surpassed again. The threshold in this approach is typically defined as the number of processor threads on the machine, adapting to different machines, but being oblivious to other factors such as memory and processor speed. In order to decrease the overhead of computing the size of queues, the size of other queues is estimated from the size of the current queue after applying a factor of (number of idle threads/active threads), because idle threads are known to have 0 tasks in their queue. This estimation assumes a regular distribution among threads, which may not always happen.

- **ATC**, or Adaptive Tasks Cut-off - This approach is a hybrid of MaxTasks and MaxLevel, changing the parallelization policy based on the recursion depth [7]. Tasks are only created if two conditions are met: there are fewer tasks than the number of threads on a parameterized recursion level; and the depth is less than a parameterized threshold. This approach forces the tasks to be created in breadth in the lower depths, and aggregated in the higher depths. The idea is to improve the speed of task distribution while preventing over-sheduling of smaller tasks in higher depths. ATC adds a profiler that saves information regarding how much time a sub-tree takes to execute, and predicts further subtrees (if the prediction is larger than 1 ms, the task will be created). This is based on the assumption that all tasks inside a level have a similar behavior, which does not happen in unbalanced programs.
- **Surplus**, or Suplus Queues Task Count - This approach is included in Java's Fork Join framework [3] and it relies on the size of work-stealing queues. Before creating a new task, the number of queued tasks in the current thread that exceeds the number of tasks in other queues is compared to a parameterized threshold limit (usually 3 in existing ForkJoin benchmarks). If the surplus tasks count is higher than the threshold, the task will be executed sequentially, meaning that the current queue already has enough work for other threads to steal. If the surplus tasks count is lower than the threshold, the task is created in parallel to create more stealable work.

In this paper, we introduce three new algorithms for performing the parallel-sequential decision for each task:

- **MaxTasksInQueue**, or Maximum Queue Size - If the current queue size is lower than a parameterized threshold, the task will be executed in parallel. If the current queue is already at its maximum capacity, tasks will be executed sequentially. This approach is similar to MaxTasks, which limits the overall number of tasks, but considers the local queue only in order to reduce the overhead in accessing information from other threads.
- **StackSize** - Many of the fine-grained irregular programs would crash from stack overflows using existing granularity algorithms. The crash would occur later in the program, much after the fully sequential version of the program would have finished. MaxLevel would consider the recursion depth of the program, but not that of the work-stealing runtime. We propose a cut-off algorithm based on the number of stack frames of the program. If the number of stack frames is lower than a parameterized value, the task will be executed in parallel.

– **MaxTasksSS**, or Maximum Tasks with Stack Size - Being based on depth, the StackSize approach is not suitable for irregular programs. In order to improve its performance, StackSize was combined with MaxTasks, resulting in a new approach. This approach uses the StackSize criteria to prevent very high granularity and uses MaxTasks criteria to allow for tasks to be created at the lower depths.

4   Cut-Off Mechanism Evaluation

In this section, we begin by introducing the experimental setup and the benchmark suite. Then, we analyze and compare the different approaches.

4.1   Experimental Environment

Two machines (Table 1) were used in order to generalize results to more than one machine, both running Ubuntu 14.04 and Java HotSpot(TM) 64-Bit Server VM with Java 1.8. Programs were implemented on top of the Æminium Runtime [10].

Table 1. Details of the hardware used in the experiments.

Name	Processor	CPU Cores	Threads	RAM
astrid	Intel Xeon E5-2650 0 @ 2.00 GHz	16 cores	32 threads	32 GB
ingrid	Intel Xeon X5660 @ 2.80 GHz	12 cores	24 threads	24 GB

To collect values, a practical statically rigorous methodology [11] was applied. For each combination of program and cut-off, we obtained a mean and the 95% confidence interval for the execution time in steady state from several executions until the Coefficient of Variance was below 5% or up to 30 executions. Each program had a timeout of 1 h. All programs were executed in the same conditions, changing only the cut-off algorithm. There was no other load on the machine besides the experiment and the operating system.

4.2   Benchmark Suite

In order to evaluate cut-off algorithms, we use a benchmark suite comprised of different fork-join programs that represent the different types of programs being written for task-based work-stealing runtimes. Table 2 shows the list of the 24 programs used, their sources and the input sizes used.

The included programs are examples of divide-and-conquer, pipelined parallelism, do-all loops, do-across loops, nested parallelism and partial parallelism in a sequential algorithm. There are balanced and unbalanced programs in the benchmark suite.

Except for do-all, all programs are real-world examples and some are used in other benchmark suites, because of their heterogeneity. Compared with other evaluations, this is the largest and most heterogeneous set of programs ever used for evaluating cut-off algorithms. The benchmark suite is freely available at https://github.com/AEminium/AeminiumBenchmarks.

Table 2. Description of the programs used in the benchmark suite

Program	Source	Input size	Type	Balance
BFS	PBBS [12]	d = 26, w = 2	Recursive	Regular
Black-Scholes	PARSEC [13]	10000^2	Loop	Regular
Convex-Hull	PARSEC [13]	10000^2	Recursive	Regular
Do-All		100 million	Loop	Regular
FFT	Cilk [14]	8388608	Recursive	Regular
Fibonacci	ForkJoin [3]	n = 47	Recursive	Irregular
Fibonacci	ForkJoin [3]	n = 49	Recursive	Irregular
Fibonacci	ForkJoin [3]	n = 51	Recursive	Irregular
Genetic Knapsack		g = 100, p = 100	Loop	Regular
Health	BOTS [15]	l = 7	Loop	Regular
Heat	ForkJoin [3]	4096×4096, it = 1024	Loop	Regular
Integrate	ForkJoin [3]	error = 10^{-9}	Recursive	Irregular
KDTree	PBBS [12]	n = 10000000	Recursive	Regular
LUD	ForkJoin [3]	4096×4096	Recursive	Regular
Matrix Mult	ForkJoin [3]	p = 10000, q = r = 1000	Loop	Regular
MergeSort	ForkJoin [3]	n = 100000000	Recursive	Regular
MolDyn	JGrande [16]	it = 1 size = 40	Loop	Regular
MolDyn	JGrande [16]	it = 5 size = 30	Loop	Regular
MonteCarlo	JGrande [16]	10000×60000	Recursive	Regular
N-Body	PBBS [12]	n = 50000, it = 3	Loop	Irregular
N-U Knapsack		items = 30, corr = 3	Recursive	Irregular
NeuralNet		it = 500000	Recursive	Regular
N-Queens	Cilk [14]	n = 8..15	Loop	Irregular
N-Queens	Cilk [14]	16	Loop	Irregular
Pi		n = 100.000.000	Loop	Regular
Quicksort	ForkJoin [3]	n = 10000000	Recursive	Regular
RayTracer	JGrande [16]	n = 2000	Loop	Regular

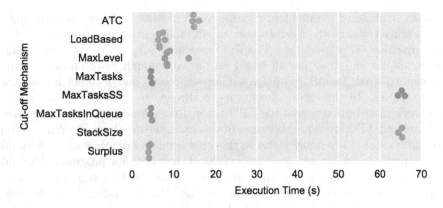

Fig. 1. Swarm plot of different cut-off approaches for the Do-all program on the *ingrid* machine.

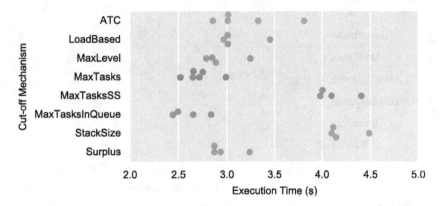

Fig. 2. Swarm plot of different cut-off approaches for the Matrix Multiplication program on the *astrid* machine.

4.3   Comparison of Cut-off Approaches

In sync with findings from prior works [6,7], this section corroborates that no cut-off approach performed better than the others for all programs. Here, the differences in performance from the algorithms are addressed. Since the time distribution of the algorithms is not normal, swarm plots will be used. For parametrized cut-off approaches, we have used the parameters that achieved the best global time in a preliminary evaluation. MaxLevel had a depth limit of 12; MaxTasks had a task limit of twice the number of threads; ATC was configured with the two limits; StackSize had a limit of 16 stacks, MaxTasksSS has the same limits as MaxTasks and StackSize; Surplus had a limit of 3 surplus task count.

Do-all is made of parallel loops with several iterations doing only one operation. Figure 1 shows the performance of different cut-off mechanisms in the
ingrid machine. MaxTasks, MaxTasksInQueue and Surplus were the most efficient strategies and they are all based on having enough work on each queue
for others to steal. LoadBased has a similar approach, but does not have extra
work in queues. In this case, allowing more threads to steal work results in a
faster work distribution across the CPU cores. Recursion-depth approaches like
MaxLevel and ATC are slower because, in this case, the depth considered was too
deep and it created too many tasks. In this case a smaller depth, such as 6 would
result in fewer tasks created, and less overhead, but in other programs it would
result in worse performances. Stack-size approaches create too many tasks as
well in this case. Figure 2 shows the same behavior in the Matrix Multiplication
program, which also has lightweight tasks in a 2 dimensional loop cycle.

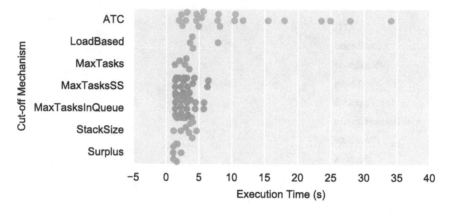

Fig. 3. Swarm plot of different cut-off approaches for the Fibonacci program on the
ingrid machine.

Figure 3 shows the Fibonacci program with different cut-offs. Fibonacci is a
highly irregular program that generates a skewed parallelization tree, with an
extremely lightweight computation. In this case all approaches handle the program reasonably well, but MaxLevel is not able to finish the program within the
defined timeout, as its execution time is not shown. Figure 4 shows integrate,
another highly irregular program, in which cut-off programs show the same relative performance, with MaxLevel being much slower than its counterparts.

In the N-Queens program, in Fig. 5, Loadbased, MaxLevel and Surplus are
the fastest algorithms. This program has a high branching factor and a high
penalty for over-creating tasks, because it needs to allocate memory for each
parallelization. MaxLevel avoids going too deep in the recursion tree, preventing
the unwanted unnecessary memory allocation. LoadBased also prevents creating
extra tasks and performs similarly to MaxLevel.

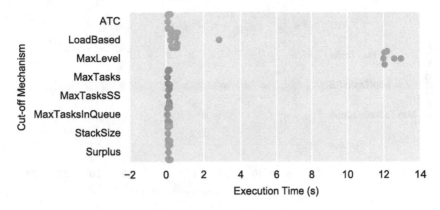

Fig. 4. Swarm plot of different cut-off approaches for the Integrate program on the *ingrid* machine.

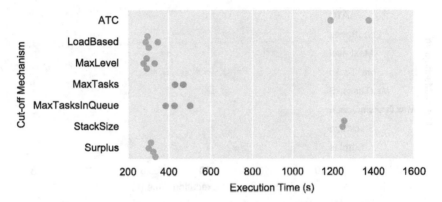

Fig. 5. Swarm plot of different cut-off approaches for the N-Queens program on the *ingrid* machine.

Figure 6 shows the cut-off performance in the FFT program. Only ATC and MaxTasksSS have finished the program 3 times within the timeout. FFT is a program that allocates a large amount of memory in its divide-and-conquer process. The allocation of tasks on top of the baseline allocation of the sequential program penalize the creation of a large number of tasks. The two best approaches have two mechanisms to limit the creation of tasks, one limiting the queue size, and another preventing from going too deep in the recursion level. The difference between the two is that ATC limits using the program recursion and MaxTasksSS uses the internal recursion of the work-stealing runtime.

In Fig. 7, we can see the opposite behavior in which ATC and MaxTasksSS are the worst approaches. One reason for this is that these hybrid approaches use two mechanisms to improve their worst-case programs, but introduce overhead in cases where the individual algorithms are ideal.

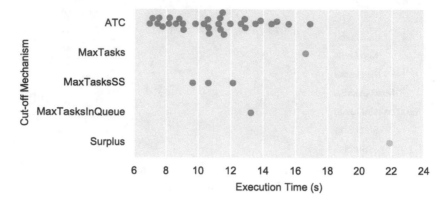

Fig. 6. Swarm plot of different cut-off approaches for the FFT program on the *ingrid* machine.

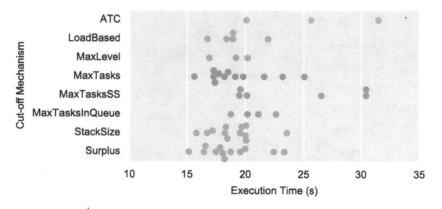

Fig. 7. Swarm plot of different cut-off approaches for the Raytracer program on the *ingrid* machine.

Figure 8 shows the same plot for the Neural Network program. In this program, creating tasks has a relatively large overhead compared to the program and only StackSize approaches have been able to complete the program within the timeout, in a relatively small time. This is one example that justifies the introduction of stack-size approaches, in which the workload of tasks is very light and there is expensive work in merging the result of each recursive call. This is the same behavior as that of the Fibonacci program with a very large input. KD-Tree is another program where Stack-based approaches are also advantageous, but not by a larger difference, which can be seen in Fig. 9.

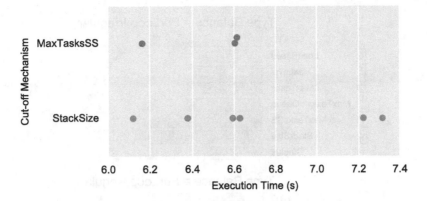

Fig. 8. Swarm plot of different cut-off approaches for the Neural Network training program on the *astrid* machine.

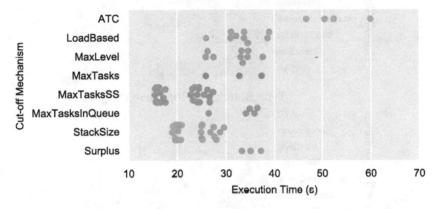

Fig. 9. Swarm plot of different cut-off approaches for the KDTree training program on the *astrid* machine.

Figure 10 shows the execution time of all the programs that are a combination of a given type and balance. While previous results were specific to each program, these plots show aggregated information from several programs. Irregular for-loop programs can be efficiently optimized using either LoadBased or MaxLevel algorithms. In general, these two algorithms do not schedule as many tasks as others because they are created either when there is an empty queue, or only in the beginning of the program. In irregular loop programs, MaxTasks is the best algorithm as it allows the creation of enough tasks to spread work across all threads, even in later iterations. Recursive irregular programs work best under Surplus because, similarly to MaxTasks, it allows for extra work to be scheduled, which improves the distribution of irregular work across threads. Finally, recursive balanced programs perform well under StackSize, which prevents high levels of recursion, regardless of the nature of the algorithm.

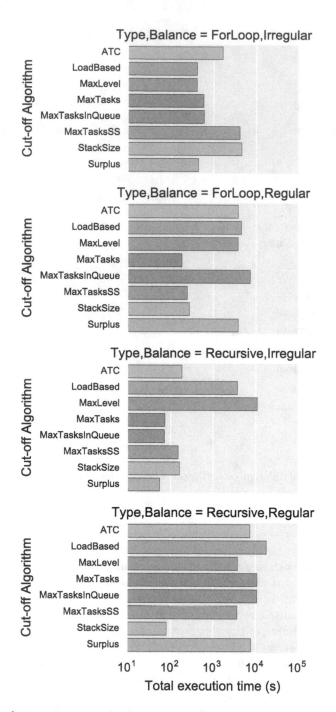

Fig. 10. Total execution time of programs that fit a certain Type-Balance pattern on the *astrid* machine.

5 Conclusions and Future Work

In this paper we have introduced three new algorithms for dynamically managing the granularity of parallel programs. Additionally, we have evaluated new and existing cut-off algorithms over a 24-program benchmark suite. The three proposed algorithms were able to outperform existing algorithms in at least one of the programs in the benchmark suite.

We have identified MaxTasks as a reasonable cut-off algorithm for a large set of programs. In irregular loop programs, Load-based and MaxLevel can be used to improve the performance of programs. Additionally, the proposed Stack-Size algorithm can be used in regular recursive programs or when the memory allocation required for parallelization is high.

For future work, we intend to analyze the structure of the source code to infer the type of parallelism and use machine-learning techniques to predict the best cut-off mechanism.

Acknowledgments. This work was partially supported by the Portuguese Research Agency FCT, through CISUC (R&D Unit 326/97), the CMU|Portugal program (R&D Project Aeminium CMU-PT/SE/0038/2008). The first author was also supported by the Portuguese National Foundation for Science and Technology (FCT) through a Doctoral Grant (SFRH/BD/84448/2012).

References

1. Dagum, L., Menon, R.: OpenMP: an industry standard API for shared-memory programming. IEEE Comput. Sci. Eng. **5**(1), 46–55 (1998)
2. Blumofe, R.D., Joerg, C.F., Kuszmaul, B.C., Leiserson, C.E., Randall, K.H., Zhou, Y.: Cilk: an efficient multithreaded runtime system, vol. 30. ACM (1995)
3. Lea, D.: A java fork/join framework. In: Proceedings of the ACM 2000 Conference on Java Grande, pp. 36–43. ACM (2000)
4. Haghighat, M.R., Polychronopoulos, C.D.: Symbolic analysis: a basis for parallelization, optimization, and scheduling of programs. In: Banerjee, U., Gelernter, D., Nicolau, A., Padua, D. (eds.) LCPC 1993. LNCS, vol. 768, pp. 567–585. Springer, Heidelberg (1994). doi:10.1007/3-540-57659-2_32
5. Mohr, E., Kranz, D., Halstead, R.: Lazy task creation: a technique for increasing the granularity of parallel programs. IEEE Trans. Parallel Distrib. Syst. **2**(3), 264–280 (1991)
6. Duran, A., Corbal, J., Ayguad, E.: Evaluation of OpenMP Task Scheduling Strategies, pp. 100–110 (2008)
7. Duran, A., Corbalán, J., Ayguadé, E.: An adaptive cut-off for task parallelism. In: Proceedings of the 2008 ACM/IEEE conference on Supercomputing, p. 36. IEEE Press (2008)
8. Olivier, S.L., Prins, J.F.: Evaluating OpenMP 3.0 run time systems on unbalanced task graphs. In: Müller, M.S., Supinski, B.R., Chapman, B.M. (eds.) IWOMP 2009. LNCS, vol. 5568, pp. 63–78. Springer, Heidelberg (2009). doi:10.1007/978-3-642-02303-3_6
9. Olivier, S.L., Prins, J.F.: Comparison of OpenMP 3.0 and other task parallel frameworks on unbalanced task graphs. Int. J. Parallel Prog. **38**(5–6), 341–360 (2010)

10. Stork, S., Naden, K., Sunshine, J., Mohr, M., Fonseca, A., Marques, P., Aldrich, J.: Æminium: a permission-based concurrent-by-default programming language approach. ACM Trans. Program. Lang. Syst. (TOPLAS) **36**(1), 2 (2014)
11. Georges, A., Buytaert, D., Eeckhout, L.: Statistically rigorous java performance evaluation. ACM SIGPLAN Notices **42**(10), 57–76 (2007)
12. Shun, J., Blelloch, G.E., Fineman, J.T., Gibbons, P.B., Kyrola, A., Simhadri, H.V., Tangwongsan, K.: Brief announcement: the problem based benchmark suite. In: Proceedings of the 24th ACM Symposium on Parallelism in Algorithms and Architectures, pp. 68–70. ACM (2012)
13. Bienia, C.: Benchmarking modern multiprocessors. PhD thesis, Princeton University, January 2011
14. Frigo, M., Leiserson, C.E., Randall, K.H.: The implementation of the cilk-5 multithreaded language. In: ACM Sigplan Notices, vol. 33, pp. 212–223. ACM (1998)
15. Duran, A., Teruel, X., Ferrer, R., Martorell, X., Ayguadé, E.: Barcelona OpenMP tasks suite: a set of benchmarks targeting the exploitation of task parallelism in OpenMP. In: 38th International Conference on Parallel Processing, pp. 124–131 (2009)
16. Smith, L.A., Bull, J.M., Obdrizalek, J.: A parallel java grande benchmark suite. In: Supercomputing, ACM/IEEE 2001 Conference, p. 6. IEEE (2001)

Implementation and Evaluation of NAS Parallel CG Benchmark on GPU Cluster with Proprietary Interconnect TCA

Kazuya Matsumoto[1]($\boxtimes$), Norihisa Fujita[2], Toshihiro Hanawa[3], and Taisuke Boku[1,2]

[1] Center for Computational Sciences, University of Tsukuba, Tsukuba, Japan
matsumoto.kazuya@jaea.go.jp
[2] Graduate School of Systems and Information Engineering, University of Tsukuba, Tsukuba, Japan
[3] Information Technology Center, The University of Tokyo, Tokyo, Japan

Abstract. We have been developing a proprietary interconnect technology called Tightly Coupled Accelerators (TCA) architecture to improve communication latency and bandwidth between accelerators (GPUs) over different nodes. This paper presents a Conjugate Gradient (CG) benchmark implementation using the TCA and results of performance evaluation on the HA-PACS/TCA system, which is a proof-of-concept GPU cluster based on the TCA concept. The implementation is based on the CG benchmark in NAS Parallel Benchmarks, and its parallelization is achieved by a two-dimensional decomposition of matrix data. The TCA utilization improves the communication performance compared with the implementation with MPI/InfiniBand utilization for small size benchmark classes. This study also shows that the CG implementation with the two dimensional decomposition is more suitable for the TCA utilization than a CG implementation with a one-dimensional decomposition to make use of the interconnect.

1 Introduction

Currently, GPU clusters are widely used as high performance computing systems. A problem of GPU clusters is that the communication speed between multiple compute nodes is not fast enough compared to its high computation speed. In order to address this problem, we have been researching the Tightly Coupled Accelerators (TCA) architecture [4]. The TCA is a technology on a proprietary interconnect network to enable direct communication between accelerators over different nodes.

We have conducted the basic performance evaluation of the TCA [3,4,7]. In [7], a Conjugate Gradient (CG) method has been implemented by utilizing allgather and allreduce collective communications with TCA's communication functions. The parallelization of the CG implementation is accomplished by a one-dimensional decomposition of matrix data. Results of the performance evaluation shows that the CG method implementation using TCA outperforms the

© Springer International Publishing AG 2017
I. Dutra et al. (Eds.): VECPAR 2016, LNCS 10150, pp. 135–145, 2017.
DOI: 10.1007/978-3-319-61982-8_14

implementation using MPI/InfiniBand for sparse matrices whose matrix (problem) size is nine thousand or smaller; however, the communication with TCA becomes less advantageous when either or both of the target matrix sizes and the number of utilizing processes are larger.

In the present study, we evaluate a different CG method implementation with the TCA utilization. The CG implementation is based on the CG benchmark in NAS Parallel Benchmarks [1]. We apply a two-dimensional decomposition of matrix as the enhanced implementation from [7], and present results of performance evaluation on the HA-PACS/TCA GPU cluster. Additionally, this study presents communication performance differences between the CG implementation with the two-dimensional decomposition and the one-dimensional decomposition.

2   Tightly Coupled Accelerators Architecture

This section briefly explains the Tightly Coupled Accelerators (TCA) architecture (see [3–5] for detailed information on the TCA). The TCA is a novel technology of proprietary interconnect for PC clusters. The PCI Express Adaptive Communication Hub ver. 2 (PEACH2) is a prototype implementation of the TCA architecture. We can construct a cluster system by connecting the PEACH2 boards with each other. The PEACH2 realizes a direct data communication between GPUs beyond nodes by utilizing the GPUDirect Support for RDMA (GDR) technology [8]. The GDR eliminates unnecessary system memory copies, lowers CPU overhead, and reduces communication latency, resulting in performance improvements on a GPU cluster. Moreover, the communication using the PEACH2 is conducted only with the PCIe protocol, and the overhead time for protocol conversion, which is required in the InfiniBand, is eliminated. Consequently, the PEACH2 enables data communication with extremely low latency. The PEACH2 provides two types of data communication functions: PIO and DMA. In the PIO communication, the data is transferred to a remote node by the CPU's remote write operation. The latency of PIO is very small, and, as a result, the PIO is useful to transfer short messages. The DMA function is achieved by the DMA controller, which has four DMA channels. While the latency of DMA is larger than that of PIO, the DMA demonstrates higher maximum bandwidth performance.

The HA-PACS (Highly Accelerated Parallel Advanced system for Computational Sciences) is a GPU cluster system at the Center for Computational Sciences, University of Tsukuba. The HA-PACS/TCA is a proof-of-concept system of TCA architecture concept and a performance evaluation test-bed of PEACH2 board. Every compute node of HA-PACS/TCA contains the PEACH2 as its interconnect adapter. Table 1 shows the specification of HA-PACS/TCA, and Fig. 1 depicts the block diagram of a compute node on HA-PACS/TCA. The HA-PACS/TCA consists of four sub-clusters. Each sub-cluster is composed of 16 compute nodes. The 16 nodes are connected by the PEACH2 and configure a 2×8 torus network. Note that the 64 nodes of HA-PACS/TCA are connected also by two ports of InfiniBand QDR in a fat-tree configuration with full bisection bandwidth.

Table 1. Node configuration and system configuration of HA-PACS/TCA

Node configuration	
Motherboard	SuperMicro X9DRG-QF
CPU	Intel Xeon E5-2680 v2 2.8 GHz × 2 (IvyBridge 10 cores/CPU)
Memory	DDR3 1866 MHz × 4 ch, 128 GB (=8 × 16 GB)
Peak performance	224 Gflops/CPU
GPU	NVIDIA Tesla K20X 732 MHz × 4 (Kepler GK110 2688 cores/GPU)
Memory	GDDR5 6 GB/GPU
Peak performance	1.31 Tflops/GPU
Interconnect	InfiniBand: Mellanox Connect-X3 Dual-port QDR TCA: PEACH2 board (Altera Stratix-IV GX 530 FPGA)
System configuration	
Number of nodes	64
Interconnect	InfiniBand QDR 108 ports switch × 2 ch
Peak performance	364 Tflops

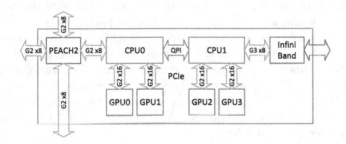

Fig. 1. Block diagram of compute node in HA-PACS/TCA

3 Implementation

The CG benchmark in NAS Parallel Benchmarks (NPB) [1] is a well-known benchmark. It measures the performance of implementation for a conjugate gradient method that conducts typical computations on unstructured grid, and tests internode communications, employing sparse matrix-vector multiplication. The performance of the NPB on GPUs has been studied to evaluate different programming models or different approaches for performance optimizations [2,6,11]. Lee and Vetter [6] compared the performance differences among different programming models such as OpenACC, HMPP, and CUDA on a single GPU. They ported and optimized the OpenMP version of NPB program. Grewe et al. [2] showed an approach that automatically generates OpenCL codes, from OpenMP programs, optimized for GPUs and the NPB are used for the performance evaluation.

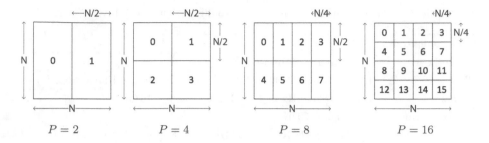

Fig. 2. Process distribution by two-dimensional decomposition of matrix data. The number in each rectangular represents its process rank id.

Xu et al. [11] studied the effectiveness of a directive based programming model of OpenACC for parallelizing NPB on GPUs.

In the present study, we modify the MPI version of CG benchmark in NPB 3.3.1 such that the main computation part (conj_grad function) is written in C language and CUDA[1], and its inter-node data communications are conducted with the TCA/PEACH2. Let us denote a linear equation system as $Ax = b$, where A is an $N \times N$ symmetric positive definite matrix, and both x and b are a vector with N elements in the following. The modified implementation uses the identical data distribution and communication pattern to the original MPI version. The parallelization of CG benchmark is achieved by a two-dimensional decomposition of the matrix A and conformable distribution of vectors. When we define the number of processes as P, the matrix A is two-dimensionally decomposed by $P = P_r \times P_c$ processes. Based on the decomposition, each process contains $(N/P_r) \times (N/P_c)$ sub-matrix of A and vectors with N/P_c elements. Figure 2 shows the process distribution by two-dimensional decomposition of matrix A data for $P = 2, 4, 8, 16$.

Almost all of computations in the implementation are carried out by GPUs. While we implement the sparse-matrix vector multiplication (SpMV) by ourselves, our CG implementation utilizes the NVIDIA's CUBLAS library for vector dot product (DOT) and vector addition (AXPY) operations. Note that the computation part is not tuned so deeply.

Three kinds of data communication are required in the implementation. The first required communication is to send vector data for obtaining the product of SpMV after the local SpMV computation on a GPU in each node. The communication is conducted among P_c processes in the same process row in a binary tree fashion, and thus $\sqrt{P_c}$ communication steps are needed (each step needs to send $8N/P_c$ Bytes of vector data in double precision). The second communication is to send a scalar (8 Bytes) value to compute the sum of local product by the DOT computation. This communication is also made among the P_c processes and $\sqrt{P_c}$ steps are required. The third communication is to send $8N/P_c$ Bytes of a vector for data exchange. In the following, let us name the first, second

---

[1] The CG benchmark program is originally written in Fortran.

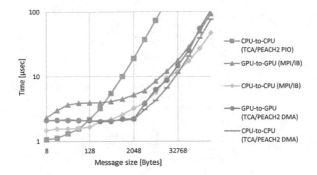

Fig. 3. Ping-pong communication performance between two neighboring nodes on the HA-PACS/TCA.

and third communication as COMM_SpMV, COMM_DOT and COMM_EXCH, respectively.

The COMM_DOT is scalar data communications between CPU memories of different processes and the latency for issuing communication operations occupies almost all of its communication time. The COMM_SpMV and COMM_EXCH are block data communications of $8N/P_c$ Bytes between different GPU memories and a communication bandwidth is important for high performance communication as well as the issue latency. Considering these communication characteristics, we implement the COMM_SpMV and COMM_EXCH with the DMA communication function of TCA/PEACH2 and implement the COMM_DOT with the PIO function. Note that, as shown in Fig. 3, the DMA communication is faster than the PIO communication when message sizes are larger than 128 Bytes (this message size is smaller than sizes for the required block communications in any problem classes of CG benchmark).

The TCA/PEACH2 configures a 2×8 torus network on a sub-cluster of HA-PACS/TCA; thus, a way of process (node) mapping also affects the communication performance. As can be seen from Fig. 2, the CG implementation with two-dimensional decomposition requires communication among P_c processes (4 processes at most when we utilize up to 16 nodes). We use a node mapping shown in Fig. 4. This mapping does not cause message data contentions and collisions within the TCA/PEACH2's communication network on a sub-cluster even when $P = 16$ cases.

4   Performance Evaluation

We conduct performance measurements on a sub-cluster of the HA-PACS/TCA. A single GPU and a single CPU are utilized per node[2]. For comparison with the

---

[2] This is because using two or more sub-clusters entails a hybrid utilization of the TCA/PEACH2 and MPI/IB, and because two or more GPUs usage requires additional considerations to use the TCA/PEACH2 effectively. Both of the hybrid utilization and the multi GPU usage are our future work.

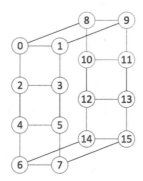

Fig. 4. Node mapping optimized for CG benchmark implementation on a sub-cluster of HA-PACS/TCA. The circles represent compute nodes, the lines between circles represent data links, and the number in each circle corresponds to its process rank id.

implementation using TCA/PEACH2, this section also presents the performance of an implementation using MPI/InfiniBand (MPI/IB) for inter-node communications. We use the MVAPICH2 GDR 2.1a (MV2GDR) [9] as MPI library implementation. As with the TCA/PEACH2, the MV2GDR utilizes the GPUDirect for RDMA (GDR) technology [8] for direct communication between GPUs. The theoretical peak bandwidth of TCA/PEACH2 (PCIe Gen2 x8) is twice lower than that of MPI/IB (dual-rail InfiniBand QDR[3]); therefore, the implementation using TCA/PEACH2 is outperformed when message sizes become large. On the condition where the program is compiled by Intel C compiler 15.0.2 with MV2GDR 2.1a and CUDA 6.5 usage, the GPU-to-GPU communication with TCA/PEACH2 is faster for message sizes up to 64 KB than the MPI/IB in terms of the ping-pong communication performance as shown in Fig. 3.

In the CG benchmark, the problem sizes (CLASS) and the number of processes (P) can be designated. This section presents results of performance evaluations for CLASS=S, W, A, B and $P = 2, 4, 8, 16$. The problem (matrix/vector) size N is 1,400 for CLASS=S, 7,000 for CLASS=W, 14,000 for CLASS=A, and 75,000 for CLASS=B. We measure the time consumed for each computation/communication operation in the conj_grad program function. Figure 5 shows the measured time breakdown on average time of ten times calls to the function. Note that this is the breakdown of process rank 0 and the communication time is the communication wait time. A single call of the conj_grad function includes $26\sqrt{P_c}$ of COMM_SpMV, $52\sqrt{P_c}$ of COMM_DOT, 26 of COMM_EXCH, 26 of SpMV, 52 of DOT, and $76 + 26\sqrt{P_c}$ of AXPY operations. In Fig. 5, the performance is shown for the implementation in Fortran with MPI/IB communication (original NPB-MPI code), C language with MPI/IB communication, CUDA with TCA/PEACH2 communication, and CUDA with MPI/IB communication.

---

[3] The theoretical peak bandwidth of the dual-rail InfiniBand QDR is 8 GB/s, which is equivalent to that of PCIe Gen3 x8.

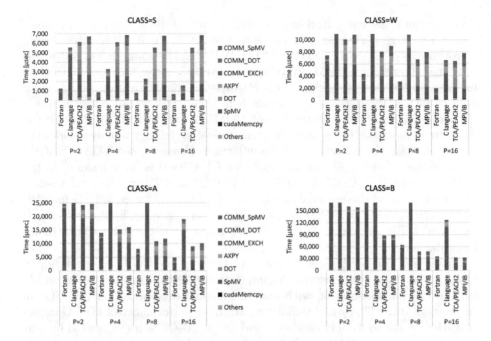

Fig. 5. Time breakdown on a single `conj_grad` function call of CG benchmark. In each figure, the performance (time in microsecond) is shown for the implementation in Fortran (original NPB-MPI code), C language, CUDA with TCA/PEACH2 communication, and CUDA with MPI/IB communication. The upper plot results for several results in Fortran and C, such as CLASS=B & P = 2 case, are cut for simplicity.

The Fortran implementation is faster than the C implementation[4]. Compared with the original Fortran implementation, the CUDA/GPU usage for NPB implementation deteriorates the performance in several cases (particularly in cases of smaller size class and larger number of processes utilization). This is mainly due to that the GPU usage brings bigger overheads for invoking the CUDA kernels and for issuing the inter-node communications between GPUs. For instance, an invocation of a simple CUDA kernel at least takes $9.7\,\mu s$, including the CUDA stream synchronization time, in our measurement. In the case CLASS=W with $P = 16$, the implementation in CUDA with MPI/IB GPU-to-GPU communication takes 6.75 times longer time for DOT communication and 2.43 times longer time for COMM_SpMV than the implementation in C with MPI/IB CPU-to-CPU communication.

The TCA/PEACH2 utilization contributes the performance improvement for the three small size classes (CLASS=S, CLASS=W, and CLASS=A) compared with the MPI/IB utilization. Especially, the communication performance is 3.00 times higher in the case CLASS=S with $P = 16$, and the overall performance

---

[4] A similar performance results were reported for the NPB-LU benchmark by Pennycook et al. [10]).

including both the computation time and communication time is 1.24 times higher. The large performance improvement by TCA/PEACH2 is derived from improvements of COMM_SpMV and COMM_DOT. The communication time is 2.42 times shorter for COMM_SpMV and 4.72 times shorter for COMM_DOT in this case. In the case CLASS=A with $P = 16$, the communication performance using the TCA/PEACH2 is 1.44 times higher and the overall performance is 1.12 times higher. The performance of COMM_DOT with TCA/PEACH2 is higher in all four class cases since the communication latency mostly decides the performance for the scalar data communication. However, the TCA/PEACH2 is not always effective. The performance for CLASS=B is almost same. In the CLASS=B case, the COMM_SpMV and COMM_EXCH are the performance bottleneck because the message size of each communication is 300 KB for $P = 2, 4$ and 150 KB for $P = 8, 16$, which are bigger than the upper limit size of 64 KB (discussed in Fig. 3); hence, TCA/PEACH2 is not advantageous to MPI/IB for the problem class.

To see how the performance is different between the CG benchmark implementation with two-dimensional decomposition (2D-CG) in the present study and the CG implementation with one-dimensional decomposition (1D-CG) in our previous study [7], we additionally measure the performance of 1D-CG implementation on the equivalent condition and cases (1D-CG is implemented by ourselves and not in the NAS Parallel Benchmarks). Figure 6 shows the measured time breakdown of the 1D-CG. The 1D-CG utilizes allgather and allreduce collective communications (see [7] for implementation details). All P processes are involved for both the collective communications in 1D-CG, whereas P_c processes

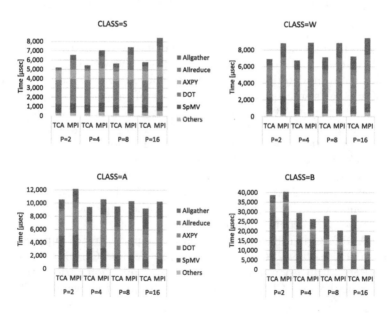

Fig. 6. Time breakdown of CG implementation with one-dimensional decomposition

are involved at most in 2D-CG. Since the network topology of TCA/PEACH2 is 2×8 torus network, collisions within the communication network cannot be avoided for $P = 16$ cases in 1D-CG. As shown in Fig. 6, the communication time in 1D-CG becomes larger when P increases. In addition, the largest message size of 1D-CG is larger than that of 2D-CG for $P = 8, 16$ cases (the size is $8N/2$ Bytes in 1D-CG and $8N/P_c$ in 2D-CG). This fact is disadvantage for communications on large matrices (especially for CLASS=B) and relative performance differences between TCA/PEACH2 and MPI/IB is large compared with 2D-CG implementation.

In general, the message size of each communication in 2D-CG is shorter than or equal to that in 1D-CG for corresponding communication. The performance degradation with shorter message size is serious in MPI/IB while TCA/PEACH2 provides a good performance thanks to its very small latency. Thus, the combination of such short messages and avoidance of message collision on the torus network of TCA/PEACH2 leads this performance improvement on 2D-CG benchmark. Figure 7 shows the strong scaling performance of 2D-CG

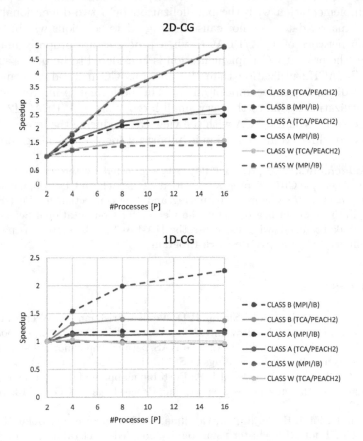

Fig. 7. Strong scaling performance of 2D-CG and 1D-CG implementations

and 1D-CG implementations. The 2D-CG implementation with TCA/PEACH2 scales better than that with MPI/IB while the 1D-CG implementation with TCA/PEACH2 scales worse.

The 2D-CG takes more overall time than 1D-CG in several cases, mostly because the 2D-CG takes around four times more time for the SpMV computation. While the SpMV of 1D-CG is conducted with the cusparseDcsrmv routine of NVIDIA's cuSparse library, the SpMV of 2D-CG is performed with a routine developed by ourselves and is less tuned. We expect that the performance of 2D-CG is equal to or higher than that of 1D-CG if the SpMV routine is tuned in a comparable level.

5   Conclusion

The present study has utilized the TCA/PEACH2 for an implementation of NAS Parallel CG benchmark and conducted its performance evaluation on the HA-PACS/TCA GPU cluster. Results of the performance evaluation show that the CG implementation with the parallelization by a two-dimensional decomposition of matrix data does not cause message data collisions within the communication network of TCA/PEACH when processes are properly mapped to nodes; and the present CG implementation is considered to be better suited for the TCA/PEACH2 utilization than the previous CG method implementation with a one-dimensional decomposition [7]. The performance improvement over MPI/IB utilization is due to the very small latency of TCA/PEACH2. We will continue researches on the TCA with the view that reducing the latency between accelerators by direct communication is important for strong-scaling computing.

Acknowledgements. The present study was supported by the Japan Science and Technology Agency's CREST program entitled "Research and Development of Unified Environment on Accelerated Computing and Interconnection for Post-Petascale Era." The authors would like to thank the Center for Computational Sciences, University of Tsukuba for allowing us to use the HA-PACS/TCA system as part of the interdisciplinary Collaborative Research Program.

References

1. Bailey, D.H., Schreiber, R.S., Simon, H.D., Venkatakrishnan, V., Weeratunga, S.K., Barszcz, E., Barton, J.T., Browning, D.S., Carter, R.L., Dagum, L., Fatoohi, R.A., Frederickson, P.O., Lasinski, T.A.: The NAS parallel benchmarks - summary and preliminary results. In: Proceedings of SC 1991, pp. 158–165 (1991)
2. Grewe, D., Wang, Z., O'Boyle, M.F.P.: Portable mapping of data parallel programs to OpenCL for heterogeneous systems. In: Proceedings of CGO 2013, pp. 1–10. IEEE (2013)
3. Hanawa, T., Fujii, H., Fujita, N., Odajima, T., Matsumoto, K., Boku, T.: Evaluation of FFT for GPU cluster using tightly coupled accelerators architecture. In: Proceedings of Cluster 2015, pp. 635–641. IEEE (2015)

4. Hanawa, T., Kodama, Y., Boku, T., Sato, M.: Tightly coupled accelerators architecture for minimizing communication latency among accelerators. In: Proceedings of IPDPSW 2013, pp. 1030–1039. IEEE (2013)
5. Kodama, Y., Hanawa, T., Boku, T., Sato, M.: PEACH2: an FPGA-based PCIe network device for tightly coupled accelerators. ACM SIGARCH Comput. Architect. News **42**(4), 3–8 (2014)
6. Lee, S., Vetter, J.S.: Early evaluation of directive-based GPU programming models for productive exascale computing. In: Proceedings of SC 2012 (2012)
7. Matsumoto, K., Hanawa, T., Kodama, Y., Fujii, H., Boku, T.: Implementation of CG method on GPU cluster with proprietary interconnect TCA for GPU direct communication. In: Proceedings of IPDPSW 2015, pp. 647–655. IEEE (2015)
8. NVIDIA: NVIDIA GPUDirect. https://developer.nvidia.com/gpudirect. Accessed 25 Aug 2016
9. Panda, D.K.: MVAPICH2-GDR (MVAPICH2 with GPUDirect RDMA). http://mvapich.cse.ohio-state.edu/overview/. Accessed 25 Aug 2016
10. Pennycook, S.J., Hammond, S.D., Jarvis, S.A., Mudalige, G.R.: Performance analysis of a hybrid MPI/CUDA implementation of the NAS-LU benchmark. SIGMETRICS Perform. Eval. Rev. **38**(4), 23–29 (2011)
11. Xu, R., Tian, X., Chandrasekaran, S., Yan, Y., Chapman, B.: NAS parallel benchmarks for GPGPUs using a directive-based programming model. In: Brodman, J., Tu, P. (eds.) LCPC 2014. LNCS, vol. 8967, pp. 67–81. Springer, Cham (2015). doi:10.1007/978-3-319-17473-0_5

Low Level Support

The Design of Advanced Communication to Reduce Memory Usage for Exa-scale Systems

Shinji Sumimoto[1]([✉]), Yuichiro Ajima[1], Kazushige Saga[1], Takafumi Nose[1], Naoyuki Shida[1], and Takeshi Nanri[2]

[1] Fujitsu Ltd., 4-1-1 Kamikodanaka 4-Chome,
Nakahara-ku, Kawasaki, Kanagawa 211-8588, Japan
{sumimoto.shinji,aji,saga.kazushige,nose.takafumi,shidax}@jp.fujitsu.com
[2] Kyushu University, 6-10-1 Hakozaki Higashi-ku, Fukuoka 812-8581, Japan
nanri@cc.kyushu-u.ac.jp

Abstract. Current MPI (Message Passing Interface) communication libraries require larger memories in proportion of the number of processes, and can not be used for exa-scale systems. This paper proposes a global memory based communication design to reduce memory usage for exa-scale communication. To realize exa-scale communication, we propose true global memory based communication primitives called Advanced Communication Primitives (ACPs). ACPs provide global address, which is able to use remote atomic memory operations on the global memory, RDMA (Remote Direct Memory Access) based remote memory copy operation, global heap allocator and global data libraries. ACPs are different from the other communication libraries because ACPs are global memory based so that house keeping memories can be distributed to other processes and programmers explicitly consider memory usage by using ACPs. The preliminary result of memory usage by ACPs is 70 MB on one million processes.

1 Motivation

Many countries have been planning to develop exa-scale systems including Japan, United States, EU and China around 2020–2023. Many core based systems will be used for the exa-scale system and the number of cores will be in the 10 million class. We have to consider not only the impacts of number of cores and nodes, but also that of the number of processes on the system software stacks in this situation, and we are researching high performance communication libraries that are able to be used for 10 million process class parallel systems.

However, current communication libraries, such as Open MPI [1] and MPICH [2], require larger memories in proportion to the number of processes, and, they can not be used for exa-scale systems because they eventually exhaust the memories. Therefore, memory usage of them must be dramatically reduced.

We propose Advanced Communication Primitives (ACPs) with global memory access and management functions to reduce memory usage by communication libraries. ACPs are aimed at achieving low-level communication primitives, and be used to implement PGAS based languages on top of ACPs.

© Springer International Publishing AG 2017
I. Dutra et al. (Eds.): VECPAR 2016, LNCS 10150, pp. 149–161, 2017.
DOI: 10.1007/978-3-319-61982-8_15

This paper is organized as follows, Sect. 2 discusses memory usage issues for Exa-scale systems. Section 3 proposes our approach of ACPs, and describes global memory based communication design to reduce memory. Section 4 shows evaluation of ACPs, and Sect. 5 discusses related work.

2   Memory Usage Issues for Exa-scale Systems

We measured and evaluated memory usage by Open MPI with an InfiniBand network using DMATP-MPI [3] tool. The evaluation environment is as follows:

- 64 node x86_64 Cluster with Quad Core AMD Opteron 8354 2.2 GHz x 4CPUs, 16 GB RAM
- Mellanox Connect X DDR InfiniBand 4HCA
- Cent OS 6.0Kernel: 2.6.32-71.29.1.el6.x86_64
- Open MPI 1.4.5.and 1.6 MVAPICH2 1.8-r5471

InfiniBand interconnect has three types of communication protocol, i.e., Reliable Connection (RC) with Receive Queue (default and RC-RQ), RC with Shared Receive Queue (RC-SRQ), and Unreliable Datagram (UD). Figure 1 plots the results.

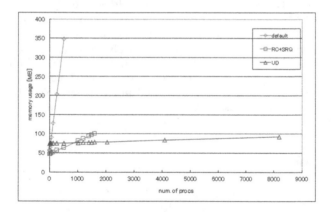

Fig. 1. Memory usage by Open MPI 1.4.5

Open MPI memory usage by the exa-scale system was estimated by extrapolating the data from Fig. 1. Table 1 summarizes the estimated memory usage by up to 10 million processes, where memory usage by RC-RQ on one million processes is 561 GB and that by RC-SRQ is 32.9 GB. These results indicate that the connection oriented communication protocol is not scalable in essentials as described by Sumimoto et al. [4].

We analyzed the reason and found that MPI_Init function allocated memory in proportion to the number of processes because each process had redundant copies of information from the other processes'. The following sub-sections describe the details.

Table 1. Estimated memory usage by Open MPI

# of node	RC-RQ	RC-SRQ	UD
100,000	56.23	3.33	0.29 GB
1,000,000	561.87	32.86	2.24 GB
10,000,000	5,618.22	328.17	21.75 GB

2.1   Evaluation of Open MPI Memory Usage and Analysis

We analyzed memory usage by the Open MPI 1.4.5[1] library by using the Dynamic Memory Allocation Tracing Profiler for MPI (DMATP-MPI) to investigate memory usage by current communication libraries [5]. The target application was IMB (Intel MPI Benchmarks). We measured several MPI functions in IMB benchmarks and chose functions that used more memory.

The results from analysis revealed that the memory usage by three MPI functions were dominant in Open MPI libraries.

- MPI_Recv: Point to Point Communication
- MPI_Init: Initializing Function
- MPI_Alltoall: Collective Communication

The following describes the results obtained from evaluating the functions, device dependent memory usage, and the memory usage of internal functions of the MPI_Init function.

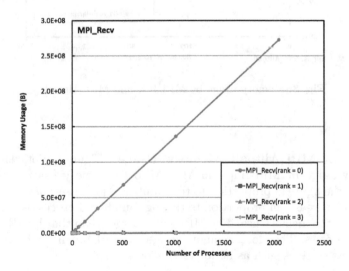

Fig. 2. Memory usage by MPI_Recv

---

[1] Current version of Open MPI is 1.8.2, however basic memory allocation is not different.

MPI_Recv Memory Usage: Figure 2 plots the results for memory usage by the MPI_Recv function on the IMB Exchange benchmark. The X axis plots the number of processes, the Y axis plots memory usage by MPI_Recv, and the results indicate memory usage from ranks 0 to 3 in the number of processes.

The usage of memory by MPI_Recv in Fig. 2 is around 140 MB in 1000 processes, and it is clear that only memory usage by rank 0 increased and the other ranks did not increase. This is because rank 0 on the benchmark played a special role that synchronized the other ranks. All ranks except for rank 0 sent messages to rank 0 to synchronize all ranks, and the messages sent to rank 0 became unexpected messages on rank 0. We also found that the unexpected messages were never freed until MPI_Finalize called, onces they were allocated.

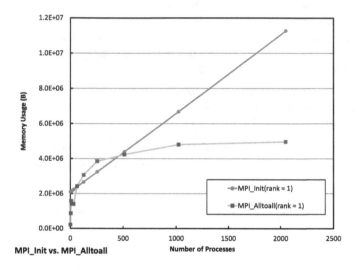

Fig. 3. Memory usage by MPI_Init and MPI_Alltoall

MPI_Init vs. MPI_Alltoall Memory Usage: Figure 3 plots the results for memory usage by MPI_Init and MPI_Alltoall. Memory usage of MPI_Init increases in the figure in proportion to the number of processes; however, that by MPI_Alltoall increases in proportion to the log (number of processes). The figure also indicates the memory usage by MPI_Init is larger than that by MPI_Alltoall over 500 processes. Therefore, memory usage by MPI_Init must be reduced in larger process systems such as node systems over 10,000.

Device Dependent Buffer Memory Usage: The result of Table 1 also indicates that memory usage of the MPI program highly depends on InfiniBand device protocol and required about 2.2 GB of memory per process even if the UD protocol was used on one million processes. This also indicates current MPI libraries usually allocate *o(Number of Processes)* amount of memory to the MPI buffer and control structure.

Followings are some description about device dependent memory usage:

InfiniBand uses Queue Pairs (QPs) which means send and receive queues to communicate with other nodes. There are two kinds of communication protocols: Un-reliable Datagram (UD) and Reliable Connection (RC). InfiniBand QP has two Work Request Queues (WRQs) to store command requests for the send queue and the receive buffer queue for the receive buffer, and the Completion Queue (CQ) for the return status of operations. RC QP requires WRQ and buffers for each destination, and UD QP requires single WRQ and buffers for all destination.

Tofu interconnect is a six dimensional torus-mesh interconnect for the K computer and the Fujitsu PRIMEHPC FX10 system. Tofu only supports RDMA data transfer and does not support message passing and it is not connection oriented. After the Tofu interconnect hardware is initialized, a program registers a memory area to a steering tag (STAG), which is the page table of Tofu, and issues RDMA operation using six dimensional coordinates. Memory usage by Tofu does not depend on the number of destinations because it does not have a message passing interface and is not connection oriented. The only dependent memories are destination lists that store node IDs and six dimensional coordinates.

Sockets are common interfaces that are widely used in current computer systems. They supports two types of communication, which are SOCK_STREAM and SOCK_DGRAM. The TCP/IP protocol is used for SOCK_STREAM, and the UDP/IP for SOCK_DGRAM. SOCK_STREAM requires socket buffer for each destination, and SOCK_DGRAM requires single socket buffer for all destinations.

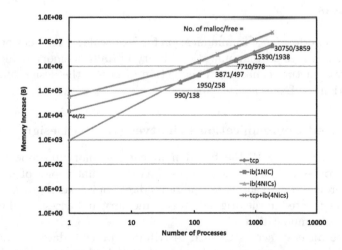

Fig. 4. Memory usage by MCA_PML_CALL(add_procs())

Table 2. Evaluation of memory allocation types in MPI_Init

Function	Device dependence	Malloc increase
MCA_PML_CALL(add_procs())	Dependent	Number of calls
ompi_proc_set_arch()	Dependent	Number of calls
ompi_proc_init()	Independent	Number of calls
MCA_PML_CALL(add_comm())	Independent	Number of calls
ompi_comm_init()	Independent	Allocation size

Internal Memory Usage by MPI_Init: We measured and evaluated several functions in MPI_Init to clarify how *o(Number of Processes)* amount of memory in MPI_Init was allocated. Figure 4 plots one of the function MCA_PML_CALL (add_procs()) results. The memory usage by Socket(tcp) and InfiniBand(ib) networks was measured, and *num1/num2* in the figure indicates the number of malloc/free functions.

The MCA_PML_CALL(add_procs()) function in Fig. 4 allocates device independent memory and the number of malloc calls increases in proportion to the number of processes.

There is a summary of the evaluation results in Table 2, where several functions in MPI_Init allocate memory in proportion to the number of processes, and allocate memory according to various characteristics, i.e., device dependent/independent and fixed size characteristics where the number of malloc calls increases or are of variable size but the number of malloc calls fixed. In any case, these functions statically have all the other process information.

3   The Design of Advanced Communication for Exa-scale Systems

To realize high performance communication for Exa-scale system, not only reduction of memory usage of communication library but also memory reduction programing infrastructure is needed. This section discusses the design of advanced communication for Exa-scale systems.

3.1   Advanced Communication Primitves (ACPs) Design

As discussed in Sect. 2, MPI_Init function allocates memory in proportion to the number of processes because each process has redundant copies of information from other processes'. To eliminate the redundant copies, such process information should be located in the original process memory and accessed when needed for exa-scale communication.

To realize an easy access to such distributed process data, we decided to introduce global memory access scheme and global addresses which is able to use an address pointer as same as address pointer in data structures in local memory. By providing the global memory access scheme, communication library users can manipulate global memory pointers without taking care of the location of the pointer.

To provide the global memory access scheme, communication library should provide functions to handle global memory access, such as data copy, data hanle, global memory allocation and global memory data library functions.

To realize the global memory access functions, we developed the Advanced Communication Primitives(ACPs). ACPs provide global addresses, which are able to use remote atomic memory operations on the global memory, and RDMA based memory copy communication to effectively manipulate distributed structure. We chose RDMA based communication because it does not need an intermediate communication buffer such as message based communication and modern interconnects such as Tofu and InfiniBand to support it.

The ACPs consist of basic layer (ACPbl) and middle layers which consist of Communication Library (ACPcl) and Data Library (ACPdl), and all of the functions are able to handle global address pointers as they are. Each process on ACPs can individually register and unregister its local memory to the global memory without inter-process synchronization. The primal data transfer function of the layer is a 'copy' on the global memory. The initiator process for the copy does not have to be the source nor the destination. The 'copy' function is directly implemented by using RDMA when network hardware, such as InfiniBand or Tofu Interconnect, has RDMA.

3.2   ACP Basic Layer: ACPbl

The basic ACP layer consists of an infrastructure, global memory management (GMM), and global memory access (GMA) functions to provide RDMA based memory copy communication, remote atomic operations, and initialize and finalize functions. It also provides fixed-size starter memory to exchange global addresses among processes after ACPs are initialized. The size of the starter memory can be used to change environment variables or argument options during execution.

ACPbl is able to provide well organized Partitioned Global Address Space (PGAS), because the starter memory is fixed sized PGAS and each node can allocate its local memory and register it as global memory. User can easily program memory space allocation as PGAS.

Current global address handles of ACPs are described as 64 bit unsigned integer type data so that they can directly use hardware atomic operation. Programs with ACPs do not have to recognize whether global address data exist on local memories or not. They only recognize them when directly accessing data. ACPs provide a translation function from global addresses to local logical memory addresses, and when the function fails, the data are on other process memories and need to be copied from the global data to local memory to access them.

Table 3. ACPbl function examples

Functions	Description	Functions	Description
acp_init()	Initialization	acp_register_memory()	Memory registration
acp_finalize()	Finalization	acp_unregister_memory()	Memory un-registration
acp_reset()	Reset	acp_copy()	Global memory copy
acp_sync()	Synchronization	acp_cas[48]()	Atomic compare and swap
acp_rank()	Getting rank number	acp_swap[48]()	Atomic swap operation
acp_procs()	Getting process group	acp_complete()	Waiting completion
acp_query	Query local address	acp_inquire()	Checking completion
_address()			

Table 3 lists examples of ACPbl functions, where there are several infrastructure functions, and copy, compare and swap, swap, checking, and waiting operation functions.

3.3   Communication and Data Libraries

ACPs are also comprised of two main categories of interfaces, i.e., communication (ACPcl) and data libraries (ACPdl [6]). The communication libraries consist of channel interface, collective and neighbor interface, and global data libraries. These interfaces are built on the basic layer that provides a global memory model among processes. Programmers create channels when needed in a channel interface, and destroy them when communication has finished. The channel interface reduces memory usage by creating and destroying channels only when needed.

ACPdl provides five types of data structures which are vector, list, deque, set and map which are similar to the collection of the C++ language standard template library. It also provides a global memory allocator function named the acp_malloc which allocates a segment of global memory from current process on a specified process rank. A global memory segment allocated by the acp_malloc function can be easily freed by the acp_free function.

4   Evaluation of ACPbl and ACPdl

We are now developing ACP libraries and have finished ACPbl for UDP/IP and Tofu and some of ACPdl to evaluate it. Fujitsu Supercomputer PRIMEHPC FX10 was used for the evaluation on Tofu interconnect, Fujitsu Supercomputer PRIMEHPC FX100 for the evaluation on Tofu2 interconnect, and Fujitsu PRIMERGY RX200 S7 for the evaluation of UDP/IP. Figure 5 plots the preliminary bandwidth performance of ACPbl for a Tofu interconnect using an acp_copy function with local memory to remote memory in Table 3.

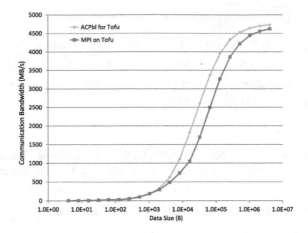

Fig. 5. Performance of ACPbl communication on Tofu

4.1 Evaluation of ACPbl Communication Performance

The performance of the communication bandwidth of MPI has also been shown for comparison, and it can be seen ACPbl outperformed MPI in bandwidth. We also evaluated preliminary memory usage by ACPbl.

4.2 Evaluation of ACPbl Memory Usage

Current estimated memory usage by ACPbl for Tofu is 70 MB on one million processes, and that for UDP/IP is 19 MB. These results are better than the results in Table 1.

Table 4. Memory usage of ACPbl(Tofu) on 1 million processes

	ACPbl(Tofu)
Memories in proportion of the # of processes – Command receive buffer – Tofu address table – Tofu routing table	69 MBytes@ 1 M processes Per process information 64 bytes/process 4 bytes/process 1 bytes/process
Memories in proportion of the # of memory registration	9 KBytes for 128 entries
Misc. buffers	262 KBytes

Table 4 shows the detail analysis of ACPbl memory usage on Tofu. In addition to the memory usage, 2 MBytes of memories are needed for Tofu hardware operation. Therefore, 72 MBytes of memories are needed for Tofu communication using ACPbl of Tofu. Table 5 shows the comparison of the memory usage

Table 5. Comparison of estimated memory usage

# of node	ACP BL Tofu	MPI RC-RQ	MPI RC-SRQ	MPI UD
100,000	0.0065	56.23	3.33	0.29 GB
1,000,000	0.0659	561.87	32.86	2.24 GB
10,000,000	0.6581	5,618.22	328.17	21.75 GB

estimation of between ACP on Tofu and MPI on InfiniBand. Memory usage of ACP on Tofu is 44.6 times less memory than MPI on UD InfiniBand.

4.3 Evaluation of ACPdl Execution Performance

This subsection presents evaluation results of ACPdl functions. In the evaluations, the acp_malloc, acp_free, acp_insert_map, and acp_find_map functions were evaluated. Every experiment used two nodes and all functions accessed memory of the other process on the remote node.

Figure 6 shows the evaluation results of the acp_malloc and acp_free functions. The average execution times of the acp_malloc and acp_free functions with the initial algorithm were around 420 and 400 usecs using the UDP version of ACPbl, 31 and 29 usecs using Tofu, and 24 and 24 usecs using Tofu2.

Figure 7 shows the results of the acp_insert_map and acp_find_map functions. The average execution times of the acp_insert_map and acp_find_map functions were around 1040 and 760 usecs using the UDP version of ACPbl, 86 and 64 usecs using Tofu, and 90 and 82 usecs using Tofu2.

These results show that ACPdl can be used effectively to handle distributed data structures with data allocation and manipulation on global memory space.

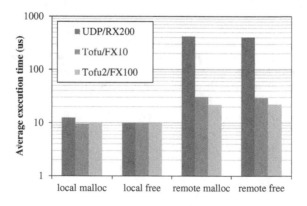

Fig. 6. Performance of ACPdl acp_malloc and acp_free

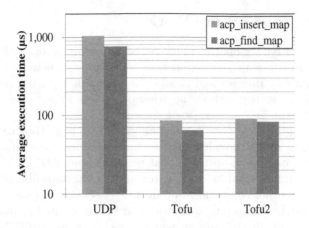

Fig. 7. Performance of ACPdl acp_insert_map and acp_find_map

5   Related Work

There have been several related work to reduce memory usage in related work.

MPICH, MVAPICH [7] and Open MPI use memory reduction techniques for InfiniBand. They use the UD and RC-SRQ communication protocol to reduce memory usage. Mellanox Dynamically Connected (DC) Transport Service, which also reduces the memory usage footprint drastically. However, DC is implemented on original InfiniBand RC protocol and some performance degradation exists when RC connections are disconnected and connected.

Open MPI only allocates a communication data buffer for communication. Open MPI for the K computer introduces two memory consumption models, i.e., high performance and memory saving modes [8]. It allocates the memory saving mode on first communication to a destination, and when the number of communications exceeds a predefined value, 16 at default, it switches the memory saving model into high performance mode.

Balaji et al. [9] discusses MPI on a million processors on MPICH and current implementation requires 80% (1.6 GB) of memory on 128 K BlueGene/P process system. It points out memory usage by communicator creation.

There are several low level communication libraries that support RDMA access and multiple networks such as UCX [10], libfabrics [11], UCCS [12], Portals [13], PAMI [14], and so on. These communication libraries provide RDMA based communication and message communication and memory usages by them depend on how to use the message communication. They do not focus memory usage reduction.

However, ACPs are true global memory based so that house keeping memories such as the other process'es information can be distributed to other processes and programmers explicitly consider memory usage by using ACPs to reduce memory usage even if for message communication.

6   Summary and Future Work

This paper proposed a global memory based communication design to reduce memory usage in exa-scale communication. We analyzed memory usage by current communication libraries and clarified issues with reducing memory usage by house-keeping memory such as the other process'es information in communication libraries.

We proposed global memory based communication primitives called ACPs to solve these issues. ACPs provide global addresses, which are able to use remote atomic memory operations on the global memory, and RDMA based memory copy communication, global heap allocator and global data libraries. ACPs are different from the other communication libraries because ACPs are global memory based so that house keeping memories can be distributed to other processes and programmers explicitly consider memory usage by using ACPs.

We have finished implementing ACPbl for UDP/IP, Tofu and InfiniBand, and ACPdl including global heap memory allocation and manipulation of five types of data structures. The preliminary evaluation results show that performance and memory usage of ACPbl outperform current MPI libraries and the preliminary result of memory usage by ACPs is 70 MB on one million processes. We intend to evaluate and optimize ACPbl and apply to several libraries such as global array, co-array and scripting languages such as python.

Acknowledgment. This research was supported by JST, CREST.

References

1. Open MPI. http://www.open-mpi.org/
2. MPICH-A Portable Implementation of MPI. http://www-unix.mcs.anl.gov/mpi/mpich/
3. Sumimoto, S., Okamoto, T., Akimoto, H., Adachi, T., Ajima, Y., Miura, K.: Dynamic Memory usage analysis of MPI libraries using DMATP-MPI. In: Proceedings of the 20th European MPI Users' Group Meeting, EuroMPI 2013, pp. 149–150. ACM (2013)
4. Sumimoto, S., Naruse, A., Kumon, K., Hosoe, K., Shimizu, T.: PM/InfiniBand-FJ: a high performance communication facility using InfiniBand for large scale PC clusters. In: Proceedings of the Seventh International Conference on High Performance Computing and Grid in Asia Pacific Region, pp. 104–113, July 2004
5. Sumimoto, S., Akimoto, H., Ajima, Y., Okamoto, T., Adachi, T., Miura, K.: Dynamic memory usage analysis of MPI libraries using DMATP-MPI. In: Proceedings of the EuroMPI 2013 (poster) (2013)
6. Ajima, Y., Nose, T., Saga, K., Shida, N., Sumimoto, S.: ACPdl: data-structure and global memory allocator library over a thin PGAS-layer. In: First International Workshop on Extreme Scale Programming Models and Middleware ESPM2 (2015)
7. MVAPICH. http://mvapich.cse.ohio-state.edu/
8. Sumimoto, S.: The MPI communication library for the K computer: its design and implementation. In: Träff, J.L., Benkner, S., Dongarra, J.J. (eds.) EuroMPI 2012. LNCS, vol. 7490, p. 11. Springer, Heidelberg (2012). doi:10.1007/978-3-642-33518-1_3

9. Balaji, P., Buntinas, D., Goodell, D., Gropp, W., Kumar, S., Lusk, E., Thakur, R., Träff, J.L.: MPI on a million processors. In: Proceedings of the 16th Euro PVM/MPI, pp. 20–30 (2009)
10. Unified Communication X. http://www.openucx.org/
11. OFI Libfabric. https://ofiwg.github.io/libfabric/
12. UCCS-Universal Common Communication Substrate. http://uccs.github.io/uccs/
13. Portals4. http://www.cs.sandia.gov/Portals/portals4.html
14. Kumar, S., Mamidala, A.R., Faraj, D.A., Smith, B., Blocksome, M., Cernohous, B., Miller, D., Parker, J., Ratterman, J., Heidelberger, P., Chen, D., Steinmacher-Burrow, B.: PAMI: a parallel active message interface for the Blue Gene/Q super-computer. In: 2012 IEEE 26th International IPDPS, pp. 763–773, May 2012

A Vectorized, Cache Efficient LLL Implementation

Artur Mariano$^{(\boxtimes)}$, Fábio Correia, and Christian Bischof

Institute for Scientific Computing, Technische Universität Darmstadt,
Darmstadt, Germany
artur.mariano@sc.tu-darmstadt.de

Abstract. This paper proposes a vectorized, cache efficient implementation of a floating-point version of the Lenstra-Lenstra-Lovász (LLL) algorithm, which is a key algorithm in many fields of computer science. We propose a re-arrangement of the data structures in LLL, which exposes parallelism and enables vectorization. We show that in one kernel, 128-bit SIMD vectorization works better than 256-bit, while in another kernel it is the other way around. In high lattice dimensions, this re-arrangement renders the implementation more cache friendly, thereby further increasing performance. Our floating-point LLL implementation is slightly slower than the implementation in the Number Theory Library (NTL) without vectorization, but 10% faster when vectorized, for lattices that require exhaustive computation with multi-precision. For larger lattices, we obtain a speedup factor of 35% over a non-vectorized implementation.

1 Introduction

Lattices are discrete subgroups of the m-dimensional Euclidean space $\mathbb{R}^m$, with a strong periodicity property. A lattice $\mathcal{L}$ generated by a basis $\mathbf{B}$, a set of linearly independent vectors $\mathbf{b}_1,...,\mathbf{b}_n$ in $\mathbb{R}^m$, is denoted by:

$$\mathcal{L}(\mathbf{B}) = \{\mathbf{x} \in \mathbb{R}^m : \mathbf{x} = \sum_{i=1}^{n} u_i\mathbf{b}_i, \mathbf{u} \in \mathbb{Z}^n\}. \tag{1}$$

where n is the rank of the lattice. When $n = m$, the lattice is said to be of full rank. When m is at least 2, each lattice has infinitely many different bases.

Lattice basis reduction is the process of transforming a given lattice basis $\mathbf{B}$ into another lattice basis $\mathbf{B}'$, whose vectors are shorter and more orthogonal than those of $\mathbf{B}$ and where $\mathbf{B}$ and $\mathbf{B}'$ generate the same lattice, i.e., $\mathcal{L}(\mathbf{B}) = \mathcal{L}(\mathbf{B}')$. While there is not a formal definition of lattice reduction, the goal of lattice reduction algorithms is to yield a *nearly orthogonal* basis.

The Lenstra-Lenstra-Lovász (LLL) algorithm was the first tractable algorithm to reduce bases [4]. It lays the foundation for many algorithms for problems on lattices. LLL has applications in many fields in computer science, ranging from integer programming to cryptanalysis [5]. Although many core concepts

© Springer International Publishing AG 2017
I. Dutra et al. (Eds.): VECPAR 2016, LNCS 10150, pp. 162–173, 2017.
DOI: 10.1007/978-3-319-61982-8_16

in the theory of lattices are already well understood, many questions regarding the performance of lattice algorithms are still under investigation. Studying the performance potential of lattice algorithms, including LLL, is of great relevance, as this determines, for example, the potential of attacks to lattice-based cryptosystems.

The original LLL algorithm was described with rational arithmetic, which was soon realized to be overly expensive. In a breakthrough result, Schnorr et al. [6] published a floating-point version of LLL that offers good practical performance and moderate stability. Since then, various improvements of LLL's stability and its algorithmic performance have been proposed e.g. [2,3,5]. For instance, in 2008, Backes et al. proposed a shared-memory parallel LLL implementation, with moderate scalability. In a follow-up paper, Backes et al. achieved an improved speedup factor of 3x for 4 threads and a bit over 4x for 8 threads [1]. However, to our knowledge, there are no studies regarding the vectorization of LLL and the impact of its data structures and kernels on cache locality.

Our contributions. In this paper, we propose a re-arrangement of the data structures in LLL that both leverages cache locality and enables SIMD vectorization. The re-arrangement of the data structures offers an immediate gain in cache locality, while the width of the SIMD vectorization should be chosen based on the pattern of computation in the kernel. We show that our floating-point LLL implementation is slower than NTL's[1], but outperforms it by 10% when employing our optimizations on bases that require multi-precision support. Moreover, the performance boost of our optimizations increases with the lattice dimension.

Notation. Vectors and matrices are written in bold face, vectors are written in lower-case, and matrices in upper-case, as in vector $\mathbf{v}$ and matrix $\mathbf{M}$. The ith coordinate of a vector $\mathbf{v}$ is denoted by $\mathbf{v}_i$. $\langle \mathbf{v}, \mathbf{p} \rangle$ denotes the inner product of two vectors $\mathbf{v}$ and $\mathbf{p}$. The Euclidean norm of $\mathbf{v}$ is given by $||\mathbf{v}||$. $\mathbf{v}$ is called a *zero vector* if $||\mathbf{v}|| = 0$. v' denotes the floating point value of an exact value v. $\lceil x \rfloor$ rounds x to the nearest integer. A detailed description of the Gram Schmidt (GS) orthogonalization, which is essential in LLL reduction, can be found in Sect. 1.2.2 of [7].

2 A LLL Floating-Point Implementation

We implemented a variant of the floating-point LLL algorithm proposed by Schnorr et al., described in [6]. Algorithm 1 shows the pseudo-code of our implementation. Floating-point LLL implementations are more practical than exact versions, but errors might occur, and a mechanism to correct them must be in place. The input of the algorithm is the lattice basis and a reduction parameter δ, which defines the extent of the reduction. The algorithm works with both *exact* and *approximate* versions of the basis. An exact copy of the basis is always available in the algorithm, since errors in the basis change the lattice and can not be corrected, unless a copy of the exact basis is kept.

---

[1] www.shoup.net/ntl/.

Algorithm 1. The heuristic LLL algorithm with floating point arithmetic, proposed by Schnorr and Euchner [6]. The lines in blue differ form the original algorithm.

Input: A basis $(\mathbf{b}_1, ..., \mathbf{b}_n)$, $\delta \in (1/2, 1)$ and all $\mu_{i,j}$'s and c_i's computed by the Gram Schmidt orthogonalization.

Output: An LLL-reduced basis with δ.

```
1   k = 2;
2   //number of precision bits in double precision
3   τ = 53;
4   Fc = false;
5   last_k = 0;
6   for i = 1, ..., m do
7       b'_i = (b_i)';

8   while k ≤ m do
9           //Computation of μ_{k,1}...μ_{k,k-1} and c_k
10          c_k = ||b'_k||²;
11          if k = 2 then
12              c_1 = ||b'_1||²;
13          for j = 1, ..., k-1 do
14              if ⟨b'_k, b'_j⟩ × ⟨b'_k, b'_j⟩ < 2^{-2×τ×0.15}||b'_k||||b'_j|| then
15                  s = ⟨b_k, b_j⟩';
16              else
17                  s = ⟨b'_k, b'_j⟩;
18              μ_{k,j} = (s - Σ_{i=1}^{j-1} μ_{j,i}μ_{k,i}c_i)/c_j;
19              c_k = c_k - μ²_{k,j}c_j;
20          //Size reduction of b_k
21          do
22              Fc = false;
23              for j=k-1, ..., 1 do
24                  if |μ_{k,j}| > 1/2 then
25                      Fc = true;
26                      for i = 1, ..., j-1 do
27                          μ_{k,i} = μ_{k,i} - ⌈μ_{k,j}⌋ × μ_{j,i};
28                      μ_{k,j} = μ_{k,j} - ⌈μ_{k,j}⌋;
29                      b_k = b_k - ⌈μ_{k,j}⌋ × b_j;
30                      if Fc then
31                          b'_k = (b_k)';
32                          if k > last_k then
33                              Recompute_GS();

34          while Fc;
35          //Swap b_{k-1}, b_k or increment k
36          if δc_{k-1} > c_k + μ²_{k,k-1}c_{k-1} then
37              swap(b_k, b_{k-1});
38              swap(b'_k, b'_{k-1});
39              if k > last_k then
40                  last_k = k;
41              k = max(k - 1, 2);
42          else
43              k = k + 1;

44  return (b_1, ..., b_n) ;
```

The algorithm starts at stage $k = 2$, by computing the Gram Schmidt orthogonalization (lines 10–19 in Algorithm 1), which starts with the computation of the inner product between two vectors. If the precision loss is too high, the exact dot product has to be computed, for which we use the exact version of the basis. The Gram Schmidt orthogonalization outputs the approximate values of one row of the coefficient vectors of the orthogonal basis, μ_k, and the square norm of the corresponding orthogonal vector.

The next step is a size reduction procedure of the vector b_k with all vectors b_j, for $j = k - 1, ..., 1$ (lines 21–34 in Algorithm 1), if the size reduction is possible. This procedure consists in subtracting the coordinates of one vector by another, whose coordinates are multiplied by a constant i.e. ($b_k = b_k - \lceil \mu_{k,j} \rfloor \times b_j$). If $|\mu_{k,j}| > 1/2$ holds true, it is possible to perform a size reduction. If the reduction takes place, we approximate the k-th row of the basis.

Finally, the reduced vector will be swapped with its predecessors unless the Lovász condition holds (lines 36–43 in Algorithm 1). This condition ensures that successive vectors are at least δ times bigger than their respective predecessor. The described process is repeated for each vector in the basis, until all vectors are LLL-reduced. Once this condition is verified, an LLL-reduced basis with δ is returned. To improve the numerical stability and the performance of the algorithm, we modified it as follows:

1. As in the NTL implementation, we replaced the 50% precision loss test of [6] by another, which tolerates a loss of up to 15% in the computation of the inner products.
2. Unlike L^3FP in [6], we check whether the values fit into a double data type (to compute the dot product in line 14 of Algorithm 1 with doubles), as we use xdoubles to store approximate values. If they do, we use doubles to compute the dot product as operations are more efficient than on xdoubles.
3. If a given basis vector b_k can be reduced, Schnorr et al. test whether the precision loss is too high. If so, the algorithm tries to reduce b_k again. However, in our implementation, b_k is always reduced again, even when the precision loss is low (lines 21–34 in Algorithm 1). This is also how the algorithm is implemented in NTL. In addition, we also recompute the Gram Schmidt orthogonalization the first time b_k is reduced, since errors may occur that are hard to correct at a later stage (lines 32–33 in Algorithm 1; note that *Recompute_GS* executes the same code as lines 10–19).

2.1   Multi-precision and Data Structures

LLL requires multi-precision capability to handle large numbers that may be present in lattices, including most lattices available from the SVP challenge[2]. A viable option to implement multi-precision is the GNU Multiple Precision Arithmetic Library (GMP) library. NTL can be compiled with either its own multi-precision module or with GMP. The LLL function in NTL is considerably

---

[2] www.latticechallenge.org/svp-challenge/.

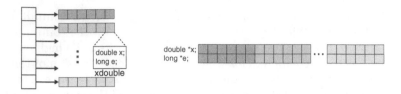

Fig. 1. Original (left side) and re-arranged (right side) data structures.

faster with its own multi-precision module than with GMP, presumably because memory can be handled much more efficiently (e.g. auxiliary variables for conversion are not needed) in the multi-precision module. In our implementation, we used GMP to store exact values.

The extended exponent double precision data type (xdouble), allows to represent floating point numbers with the same precision as a double, but with a much larger exponent. It is implemented as a class, where two instance variables are used, a double x and a long e, to store the mantissa and the exponent, respectively. For any given number in the form $x \times b^e$, x denotes the mantissa, b the base and e the exponent.

The data structures of our *base* implementation consist of 2-dimensional arrays, of either xdoubles for floating point arithmetic (GS coefficients μ and the approximate basis **B**′), or the GMP mpz_t data type for exact arithmetic (exact basis **B**), for matrices. For vectors, we used 1-dimensional arrays containing xdoubles (square norms of the GS vectors - no vectors with exact precision are needed). In addition, two xdouble arrays are used to store the square norms of the approximated basis vectors (used in line 14 of Algorithm 1) and the result of $\mu_{k,i}c_i$ (computed in line 18 and needed again in line 19 of Algorithm 1).

2.2 Data Structure Re-organization and Vectorization

We now describe two core modifications of our LLL implementation, which improve its performance. Figure 1 shows the re-arrangement of the data structure to store the approximate version of the lattice basis. On the left side, we store an array of N pointers to other arrays, each of which has N elements. Each element is stored as a xdouble object, which is a struct of two elements (a double and a long). On the right side, we show the data structure re-arranged. This re-arrangement results in immediate performance boost, as it is more cache friendly.

As the original data structures are an array of structs (AoS), cache locality is low. With the re-arrangement, multiple vectors are brought to cache with two accesses (arrays double *x and long *e). A vector in dimension N has N coordinates of 16 bytes each (8 bytes for the long and 8 bytes for the double). Therefore, accessing array *x brings 8 elements to each L1 cache line, assuming a 64 bytes L1 cache line size. This is also true for cache lines in L2, thus reducing memory access latency in comparison to the original implementation.

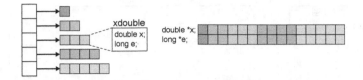

Fig. 2. Original (left side) and re-arranged (right side) data structures.

We store the GS coefficients μ in an identical data structure, although it has the shape of a lower triangular matrix. The re-arrangement is similar, as shown in Fig. 2. The major difference is index calculation. In the new format, $\mu_{i,j}$ is accessed at the index $(i \times (i-1)/2) + j$, thereby incurring in a slight overhead.

These re-arrangements also allow one to vectorize (i) the *dot product* between two vectors when they fit in doubles (line 14 in Algorithm 1) and (ii) the add and multiply (*AddMul*) (line 18 in Algorithm 1). Note that when vectors do not fit into doubles, no vectorization is used in (i), as this kernel represents a tiny percentage of the overall execution time. For (ii), we were able to partially vectorize the operation, as it is performed exclusively with xdoubles. We split the kernel in two steps. First, we multiply the elements (xdoubles) of one array by the corresponding elements of a second array, which has no dependencies and can be vectorized. In particular, the mantissas are multiplied by one another and the exponents are summed up, and both operations are vectorized. Then, we sum up the partial multiplications. However, there is a case statement in the sum which impedes vectorization.

3   Experiments

As mentioned before, we used NTL's implementation of LLL as a reference implementation. We note that NTL's implementation is faster than our base implementation due to two main reasons: (1) NTL uses its own multi-precision module, which is more efficient than GMP (which we used in our implementation), and (2) NTL's LLL implementation is more efficient than ours in terms of Gram Schmidt computations. However, our main goal is to propose optimizations that can be applied to any LLL implementation (including NTL's).

Throughout this section, we refer to our implementation as either (i) *base* implementation, for the non-optimized, implementation, (ii) *optimized/OPT*, for the version with the data structures re-arranged or (iii) *vectorized/VEC* for the version with re-arranged data structures and vectorization enabled. For 256-bit SIMD vectorization we used AVX2, while for 128-bit SIMD vectorization SSE 4.2 was used.

We used random Goldstein-Mayer lattice bases, available on the SVP challenge website, for which we ran 50 seeds on each dimension. For tests with Ajtai lattices from the Lattice challenge[3], we run tests on a single seed for each

---

[3] http://www.latticechallenge.org/.

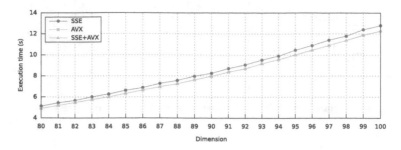

Fig. 3. Execution time of our LLL implementation, vectorized with SSE 4.2, AVX 2 and a mix of SSE 4.2 (for the *AddMul* kernel) and AVX 2 (for the *dot product* kernel) on Goldstein-Mayer lattices.

dimension, as no lattice generator is available. The test platform has two Intel E5-2698v3 chips at 2,3 GHz, each of which has 16 cores. Each core has 32 KB of L1 instruction and data cache (a cache line has 64 bytes). L2 caches have 256 KB and are not shared. The L3 cache is shared among all the cores, and has 40 MB. The machine has 756 GBs of RAM.

The code was compiled with GNU g++ 4.8.4. We compiled the code with the -O2 optimization flag, since it was slightly better than -O3.

3.1 Goldstein-Mayer Lattices (Low Dimensions)

In this section, we show the benchmarks that were carried out for lattices of the SVP challenge. We used the lattice generator to generate 50 lattices with seeds 1–50 and thus have a statistical significant result. We run our LLL implementation and NTL's implementation for lattices in dimensions 80–100. The performance of our *base* implementation is comparable to NTL's (it is about 3% slower), as shown in Fig. 4 (note the zoom-in section where the performance difference is accentuated). We did not extend the benchmarks to higher dimensions as the pattern seems to be fairly stable and higher dimensions require large chunks of time to be tested (dimension 100 required about 14 s × 50 seeds = ≈12 min, and dimension 150 would require about 3.5 h).

Figure 3 shows a comparison of our LLL implementation using different vectorization technologies. The performance of the *dot product* kernel is higher when using the better vectorization technology (AVX 2). However, this is not true for the *AddMul* kernel, whose performance is equivalent with either AVX 2 or SSE 4.2. Since this kernel computes the values of $\mu_{j,i}\mu_{k,i}c_i$, for $i = 1, ..., j-1$ and $j = k-1, ..., 1$, the number of vectorized elements is higher with SSE 4.2, thus achieving the same performance as with AVX 2.

We note that our *optimized* (OPT) version does not perform necessarily better than the *base* version. We believe that this happens because the lattices we tested are too small to exhibit enough cache locality gains to outweigh the overhead incurred in this version. To prove this, we measured the cache misses of our implementation for the L1, L2 and L3 level caches, as shown in Figs. 5,

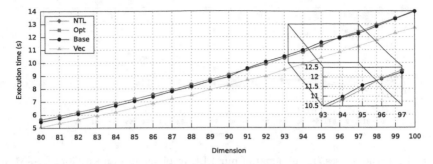

Fig. 4. Execution time of our LLL implementation and NTL's, for lattices from the SVP challenge. Note the zoom-in section for Base and OPT, between dimensions 93–97.

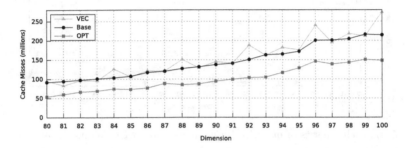

Fig. 5. L1 cache misses (in millions) of our LLL implementation on Goldstein-Mayer lattices.

6 and 7, respectively. As the figures show, the OPT version incurs much fewer cache misses across all cache levels than the base version. In particular, the difference increases with higher lattice dimensions. This shows that one can effectively improve the data locality of the code by replacing Arrays of Structures by Structures of Arrays. Ideally, we would test lattices in dimension 500–1000 but these dimensions would be impractical to solve on this type of lattices. In the next section, we test lattices in dimensions 500–800, on a kind of lattices that requires far less time.

Our *vectorized* (VEC) version obtains speedups of 9–11% over the optimized version, when using 256-bit SIMD vectorization for the *dot product* kernel and 128-bit SIMD vectorization for the *AddMul* kernel. This version incurs, somewhat surprisingly, more cache misses than the other versions. We believe that this happens because, as performance increases, more memory accesses are performed within the same timespan, thus shortening the window opportunity for efficient prefetching. In particular, for dimensions that are divisible by 4, the time for prefetching is smaller, due to higher throughput and it is not possible to load as much data beforehand, thus incurring in a higher number of L1 cache misses. Similarly, on dimensions that are multiples of 2, the VEC version also has a slightly higher number of L1 cache misses, due to the fact that some of the

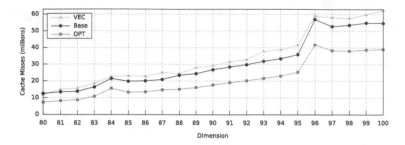

Fig. 6. L2 cache misses (in millions) of our LLL implementation on Goldstein-Mayer lattices.

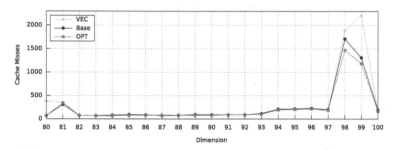

Fig. 7. L3 cache misses of our LLL implementation on Goldstein-Mayer lattices.

code is vectorized with 128-bit SIMD vectorization. For L2 and L3 cache levels, this pattern is not verified because as the latency incurred to access L2 and L3 cache levels provides enough of a time window for prefetching.

With 256-bit SIMD vectorization, we could obtain a theoretical maximum speedup of 4x (as we vectorize 8-byte doubles) and with 128-bit SIMD vectorization we could obtain a theoretical maximum speedup of 2x, for the same reason (but for 8-byte longs). Thus, in theory, we could achieve an overall speedup of 19.5%, as the *dot product* loop takes approximately 16% of the execution time of the base version (for a lattice in dimension 100), for which we used 256-bit SIMD registers, while the *AddMul* loop takes approximately 31%, for which we used 128-bit SIMD registers[4]. A 11% speedup is in our view a good result, as the maximum number of vectorized elements is N (in this case 100, at most), which is not sufficient to achieve the full potential of vectorization.

3.2   Ajtai Lattices (High Dimensions)

For lattices from the SVP challenge, LLL is only practical below dimension 200, as we mentioned in the previous subsection. The Lattice challenge allows one to carry out benchmarks with larger lattice dimensions, as this kind of

---

[4] As the number of elements that are vectorized in the loop decreases, there may not be 4 elements, which are necessary to use 256-bit SIMD.

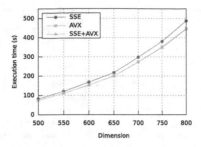

Fig. 8. Execution time of our LLL implementation, vectorized with SSE 4.2, AVX 2 and a mix of SSE 4.2 (for the *AddMul* kernel) and AVX 2 (for the *dot product* kernel) on Ajtai lattices.

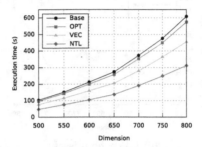

Fig. 9. Execution time of our LLL implementation and NTL's LLL implementation, for Ajtai lattices from the Lattice challenge.

lattices contain far smaller numbers and LLL-reduces them much faster. Note that lattices from the SVP challenge have numbers with over 300 digits after a certain dimension, while lattices from the Lattice challenge have numbers with no more than 3 or 4 digits.

Figure 8 shows the execution time of our LLL implementation, when both kernels are vectorized with SSE 4.2, when both kernels are vectorized with AVX 2, and when the *AddMul* kernel is vectorized with SSE 4.2 and the *dot product* kernel with AVX 2. As the figure shows, the *dot product* always benefits from AVX 2. On the other hand, the *AddMul* kernel does not profit from AVX 2 on most of the tested lattices, since the number of elements to be vectorized might not be enough to fill the 256-bit SIMD registers. Using AVX 2 on this kernel leads to a higher number of unvectorized elements, as with SSE 4.2 one can vectorize the operation as soon as the loop has 2 elements (while 4 elements are necessary for AVX 2).

Figure 9 compares our implementation against NTL's, for lattices between dimension 500 and 800. NTL is approximately 2x faster than our base implementation, as (i) NTL saves more GS computations in higher lattice dimensions and (ii) converting data types from/to GMP, which we use, incurs increasing overhead with the lattice dimension. However, the key point in this subsection is not to show how our implementation compares to NTL, but what performance

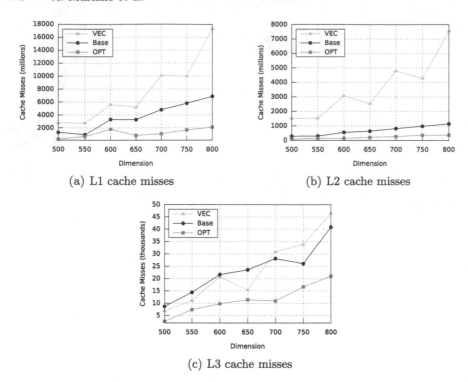

(a) L1 cache misses   (b) L2 cache misses

(c) L3 cache misses

Fig. 10. L1, L2 and L3 cache misses of our LLL implementation, for Ajtai lattices from the Lattice challenge.

gain can be attained when optimizing it. As the figure shows, we obtain a 6% speedup by simply switching to the optimized (aka with re-organized data structures) version. This backs up our claim that re-organizing the data structures delivers higher gains for higher lattice dimensions, as the experiments in the last subsection were only done for lattices up to dimension 100. In addition, Fig. 10 shows that the optimized version reduces the number of cache misses in comparison to the base version, thereby confirming that the data structure re-organization improves the data locality of the implementation.

In addition, we obtain a speedup of as much as 35% (from which 6% is obtained from the data structures re-arrangement). For the vectorization, we could achieve a theoretical speedup of 36.9%, as the *dot product* loop takes approximately 26.4% of the execution time of the base version (for a lattice in dimension 100), for which we used 256-bit SIMD registers, while the *AddMul* loop takes approximately 60.6%, for which we used 128-bit SIMD registers. The overall speedup of 35% (29%, if the speedup from the re-arrangement is deducted) is thus closer to the maximum possible speedup of 36.9%, which backs up our claim that the vectorization benefit increases with the lattice dimension.

4   Conclusions

Although a comprehensive body of work pertaining to LLL has been published in the last decades, there are no studies regarding the vectorization of LLL and the impact of its data structures and kernels on cache locality. In this paper, we fill this gap in knowledge. We propose a re-organization of the data structures in the algorithm, which enables the vectorization of two computationally expensive kernels. We show that (i) our data structure re-arrangement increases performance with the lattice dimension (ii) vectorizing the *dot product* and *AddMul* kernels can achieve as much as 35% speedup on larger lattices and (iii) our implementation is as much as 10% more efficient than NTL's on smaller lattices.

References

1. Backes, W., Wetzel, S.: Improving the parallel schnorr-euchner LLL algorithm. In: Xiang, Y., Cuzzocrea, A., Hobbs, M., Zhou, W. (eds.) ICA3PP 2011. LNCS, vol. 7016, pp. 27–39. Springer, Heidelberg (2011). doi:10.1007/978-3-642-24650-0_4
2. Koy, H., Schnorr, C.P.: Segment LLL-reduction of lattice bases. In: Silverman, J.H. (ed.) CaLC 2001. LNCS, vol. 2146, pp. 67–80. Springer, Heidelberg (2001). doi:10.1007/3-540-44670-2_7
3. Koy, H., Schnorr, C.P.: Segment LLL-reduction with floating point orthogonalization. In: Silverman, J.H. (ed.) CaLC 2001. LNCS, vol. 2146, pp. 81–96. Springer, Heidelberg (2001). doi:10.1007/3-540-44670-2_8
4. Lenstra, A., Lenstra, H., Lovász, L.: Factoring polynomials with rational coefficients. Math. Ann. **261**, 515–534 (1982)
5. Nguên, P.Q., Stehlé, D.: Floating-point LLL revisited. In: Cramer, R. (ed.) EURO-CRYPT 2005. LNCS, vol. 3494, pp. 215–233. Springer, Heidelberg (2005). doi:10.1007/11426639_13
6. Schnorr, C., et al.: Lattice basis reduction: Improved practical algorithms and solving subset sum problems. Math. Programm. **66**, 181–191 (1993)
7. Stehlé, D.: Floating-point LLL: theoretical and practical aspects. In: Nguyen, P.Q., Vallée, B. (eds.) The LLL Algorithm - Survey and Applications, pp. 179–213. Springer, Heidelberg (2010)

Versat, a Minimal Coarse-Grain Reconfigurable Array

João D. Lopes and José T. de Sousa[⊠]

INESC-ID/IST, University of Lisbon, Lisbon, Portugal
jose.desousa@inesc-id.pt

Abstract. This paper introduces Versat, a minimal Coarse-Grain Reconfigurable Array (CGRA) used as a hardware accelerator to optimize performance and power in a heterogeneous system. Compared to other works, Versat features a smaller number of functional units and a simpler controller, mainly used for reconfiguration and data transfer control. This stems from the observation that competitive acceleration can be achieved with a smaller array and more flexible reconfigurations. Partial reconfiguration plays a central role in Versat's runtime reconfiguration scheme. Results on core area, frequency, power and performance are presented and compared to other implementations.

Keywords: Reconfigurable computing · Coarse-grain reconfigurable arrays · Heterogeneous systems

1   Introduction

A suitable type of reconfigurable hardware for embedded devices is the Coarse-Grain Reconfigurable Array (CGRA) [1]. Fine grain reconfigurable fabrics, such as FPGAs, are often too large and power hungry to be used as embedded cores. It has been demonstrated that certain algorithms can run orders of magnitude faster and consume lower power in CGRAs when compared to CPUs (see for example [2]).

A CGRA is a collection of programmable functional units and embedded memories, interconnected by programmable switches for forming hardware data-paths that accelerate computations. The reconfigurable array is good for accelerating program loops with data array expressions in their bodies. However, the parts of the program which do not contain these loops must be run on a more conventional processor. For this reason, CGRA architectures normally feature a processor core. For example, the Morphosys architecture [3] uses a RISC processor and the ADRES architecture [4] uses a VLIW processor.

This work started with 3 observations: (1) because of Amdahl's law, accelerating kernels beyond a certain level does not result in significant overall acceleration and energy reduction of the application; (2) the kernels that are best accelerated in CGRAs do not require much control code by themselves; (3) the compute intensive inner loops that are normally accelerated in CGRAs tend to

© Springer International Publishing AG 2017
I. Dutra et al. (Eds.): VECPAR 2016, LNCS 10150, pp. 174–187, 2017.
DOI: 10.1007/978-3-319-61982-8_17

cluster in the code. About observation (2), note that typical target kernels are transforms (IDCT, FFT, etc.), filter banks (FIR, IIR, etc.), and others. Observation (3) refers to loop nests and loop sequences where the data produced in one loop is consumed in the next loop. If the hardware does not support these constructs, the associated processor needs to reconfigure the array after each inner loop, and the resulting time overhead may cancel the acceleration gains. Previous work has targeted this problem by proposing CGRAs that can support nested loops. For example, the approach in [5] supports nested loops using special address generation units and has been successfully used in commercial audio codec applications.

This paper extends the work in [5] to sequences of loop nests, achieving further acceleration and increasing the granularity of the tasks handled by the reconfigurable array. We propose a new architecture, Versat, which uses a relatively small functional unit array coupled with a very simple controller. A smaller array limits the size of the data expressions that can be mapped to the CGRA, forcing large expressions to be broken down into smaller expressions executed sequentially in the CGRA. Therefore, Versat requires mechanisms for handling large numbers of configurations efficiently and flexibly.

Versat cores are to be used as co-processors in an embedded system containing one or more commercial application processors. The advantage of having one or more Versat cores in the system is to optimize performance and energy consumption during the execution of compute intensive tasks. Application programmers can use the Versat cores by calling procedures that get executed on the Versat cores. For that purpose, a Versat API library must be linked with the host application. The API library is created by Versat programmers. In this way, the software and programming tools of the CGRA are clearly separated from those of the application processor.

A compiler for Versat has been developed. The compiler is simple as we have restricted its functionality to the tasks that CGRAs can do well. The syntax of the programming language is a small subset of the C/C++ language, with a semantics that enables the description of hardware datapaths. The compiler is not described in this paper whose main thrust is the description of the architecture and VLSI implementation.

In order to make the reconfiguration process efficient, full reconfiguration of the array should be avoided. In this work we exploit the similarity of different CGRA configurations by using *partial reconfiguration*. If only a few configuration bits differ between two configurations, then only those bits are changed. Most CGRAs are only fully reconfigurable [3,4,6] and do not support partial reconfiguration. The disadvantage of performing full reconfiguration is the amount of configuration data that must be kept and/or fetched from external memory. Previous CGRA architectures with support for partial reconfiguration include RaPiD [7], PACT [8] and RPU [2]. RaPiD supports dynamic (cycle by cycle) partial reconfiguration for a subset of the configuration bitstream, using smaller stored contexts. In PACT, one of its processors has access to the configuration memory of the array, but using this feature for partial reconfiguration is

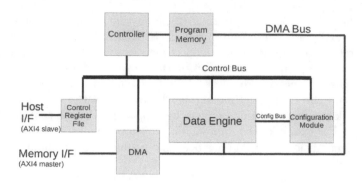

Fig. 1. Versat top-level entity.

reportedly slow and users are recommended to avoid it and resort to full reconfiguration whenever possible. In RPU, a kind of partial reconfiguration called Hierarchical Configuration Context is proposed to mitigate these problems. In this work we propose a configuration register file using registers of variable length, organized in configuration spaces and low-level configuration fields, where each register corresponds to a configuration field, and we allow random access to the configuration fields. This scheme is more flexible than the hierarchical organization of the configuration contexts in [2].

2   Architecture

The top level entity of the Versat module is shown in Fig. 1. Versat is designed to carry out computations on data arrays using its Data Engine (DE). To perform these computations the DE needs to be configured using the Configuration Module (CM). A DMA engine is used to transfer the data arrays from/to the external memory. It is also used to initially load the Versat program and to move CGRA configurations to/from external memory.

The Controller executes programs stored in the Program Memory (8 kB). A program executes an algorithm, coordinating the reconfiguration and execution of the DE and the DMA. The controller accesses the modules in the system by means of the Control Bus.

Versat has a host interface and a memory interface. The host interface is used by a host system to command the loading and execution of programs. The host and the Controller communicate using the Control Register File (CRF). The memory interface is used to access data from an external memory using the DMA.

2.1   Data Engine

The Data Engine (DE) has a fixed topology using 15 functional units (FUs) as shown in Fig. 2. It is a 32-bit architecture and contains the following FUs: 4 dual-port 8 kB embedded memories, 4 multipliers, 6 Arithmetic and Logic Units

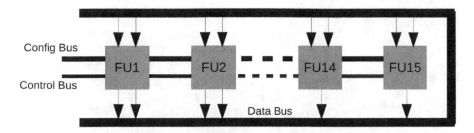

Fig. 2. Data engine.

(ALUs) and 1 barrel shifter. The Controller can read and write the output register of the FUs and can read and write to the embedded memories. In this work, embedded memory blocks are treated like any other FU by our mapping tool.

Each FU contributes its 32-bit output(s) to a wide Data Bus of 19×32 bits, and is able to select one 32-bit data bus entry for each of its inputs. The FUs read their configurations from the Config Bus. Each FU is configured with an operation and input selections. The *coarse-grain reconfiguration* means that there is a fixed set of operations available in the accelerator. For example, an ALU can be configured to perform addition, subtraction, logical AND, maximum and minimum, etc.

In Fig. 3, it is shown in detail how a particular FU is connected to the control, data and configuration busses. The FU is labeled FU5 and it is of type ALU. It is a pipelined ALU with 2 pipeline stages. The last pipeline stage stores the output of the ALU (output register). FU5 is selecting one of the 19 sub-busses of the Data Bus for each of its two inputs. Although Fig. 2 shows the Config Bus going to all FUs, in fact only the configuration bits of each FU are routed to that FU. These bits are called the *configuration space* of the FU. The configuration space is further divided in *configuration fields* with specific purposes. In Fig. 3, the example ALU has a configuration space with 3 configuration fields: the selection of the ALU's input A (5 bits), the selection of the ALU's input B (5 bits) and the selection of the ALU's function (4 bits). Our partial reconfiguration scheme works at the field level. Fields can be reconfigured one by one by the Controller. The ALU output (pipeline register 1) can be read or written by the Controller as shown in the figure. This feature enables a functional unit to be used as a shared register between the Controller and the DE.

From the explanation in the previous paragraph, one concludes there are direct connections from any FU to any other FU. This complete mesh topology may be unnecessary but it greatly simplifies the compiler design as it avoids expensive place and route algorithms commonly used in CGRAs. More compact interconnect may be developed in the future simultaneously with compiler improvements. In any case, the interconnect consumes very little power since Versat is reconfigured only after a complete program loop is executed in the DE. Moreover our IC implementation results indicate that only 4.04% of the core area is occupied by the full mesh interconnect, which means there is little

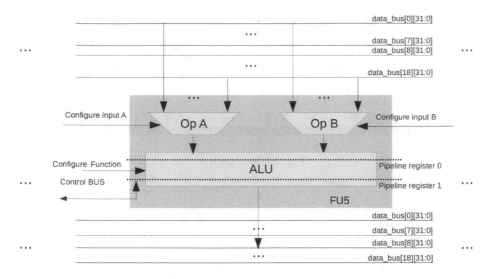

Fig. 3. Functional unit detail.

motivation to optimize the interconnect. One could argue that a full mesh topology also limits the frequency of operation. However, our IC implementation is able to work at a maximum frequency of 170 MHz in a 130 nm process, while many target applications that we have investigated, for example, in the multimedia space, are required to work at even lower frequencies because of power and energy constraints.

Each configuration of the DE corresponds to one or more hardware datapaths. Datapaths can have parallel execution lanes to exploit Data Level Parallelism (DLP) or pipelined paths to exploit Instruction Level Parallelism (ILP). Given enough resources, multiple datapaths can operate in parallel in Versat. This corresponds to having Thread Level Parallelism (TLP). In Fig. 4, three example hardware datapaths that can be mapped onto the DE are illustrated. Datapath (a) implements a pipelined vector addition. Despite the fact that a single ALU, configured as an adder, is used, ILP is being exploited: the memory reads, addition operation and memory write are being executed in parallel for consecutive elements of the vector. Datapath (b) implements a vectorized version of datapath (a) to illustrate DLP. The vectors to be added spread over memories M0 and M2, so that 2 additions can be performed in parallel. ILP and DLP can be combined to yield very parallel datapaths such as datapath (c), whose function is to compute the inner product of two vectors. Four elements are multiplied in parallel and the results enter an adder tree followed by an accumulator.

Each memory port is equipped with an Address Generator Unit (AGU) to access data from the embedded memories during the execution of a program loop. The discussion of the details of the AGU falls out of the scope of this paper. Our scheme is similar to the one described in [9] in the sense that both schemes use parallel and distributed AGUs. We will simply state the following

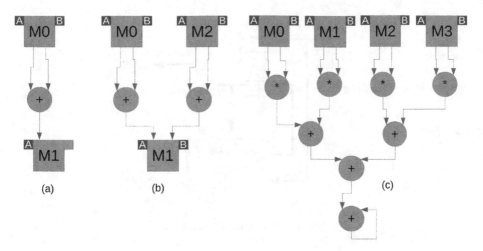

Fig. 4. Data engine datapaths.

properties of our AGUs: (1) two levels of nested loops are supported (reconfiguration after each inner loop would cause excessive reconfiguration overhead); (2) the AGUs can start execution with a programmable delay, so that paths with different accumulated latencies can be synchronized; (3) one AGU can be started independently of the other AGUs, which may be at rest or running.

The third property is instrumental for exploiting TLP, which can be illustrated using datapath (b) in Fig. 4. Suppose one block of vector elements to be added are placed in memory M0, and that address generators M0-A, M0-B and M1-A are started right away (Thread 1). In parallel, one can move the next block to memory M2 and start AGUs M2-A, M2-B and M1-B (Thread 2). Then the Controller can monitor the completion of Thread 1 in order to restart it with a new vector block, and then monitor the completion of Thread 2 to also restart it with a new block. By alternately managing Thread 1 and Thread 2, vectors that largely exceed the capacity of the Versat memories can be processed in a continuous fashion.

2.2 Configuration Module

The set of configuration bits is organized in configuration spaces, one for each FU. Each configuration space may contain several configuration fields. All configuration fields are memory mapped from the Controller point of view. Thus, the Controller is able to change a single configuration field of a functional unit by writing to the respective address. This implements partial reconfiguration. Configuring a set of FUs results in a custom datapath for a particular computation.

The Configuration Module (CM) is illustrated in Fig. 5 with a reduced number of configuration spaces and fields for simplicity. It contains a variable length configuration register file, a configuration shadow register and a configuration memory. The configuration shadow register holds the current configuration of the DE,

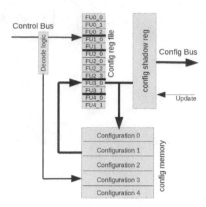

Fig. 5. Configuration module.

which is copied from the main configuration register whenever the Update signal is asserted. In this way, the configuration register can be changed while the DE is running. Figure 5 shows 5 configuration spaces, FU0 to FU4, where each FUj has configuration fields FUj_i. Note that, unlike what is suggested by the figure, the FUj_i fields do not have necessarily the same length (number of bits). A configuration memory that can hold 5 complete configurations is also shown. In the actual implementation the configuration word is 660 bits wide, there are 15 configuration spaces, 110 configuration fields in total and 64 configuration memory positions.

Still referring to Fig. 5, if the CM is being addressed by the Controller, the decode logic checks whether the configuration register file or the configuration memory is being addressed. The configuration register file accepts write requests and ignores read requests. The configuration memory interprets read and write requests as follows: a read request causes the addressed contents of the configuration memory to be read into the configuration register file; a write request causes the contents of the configuration register file to be stored into the addressed position of the configuration memory. This is a mechanism for saving and loading entire configurations in a single clock cycle with all configuration fields concatenated in a 660-bit word.

The CM has a special address that, when the Controller writes anything to it, all bits of the configuration register are cleared in one clock cycle. The default values of the configuration fields are coded with the value zero, so that this action restores the default configuration. Building a configuration of the DE from the default configuration is about 40% faster than writing all fields because many fields are left with their default values. The default values have been chosen so that they have a high likelihood of being used.

In most applications there is also a high likelihood that one configuration will be reused again, as is or with little modifications. Thus, it is useful to save certain configurations in the configuration memory to later load them and eventually tweak them.

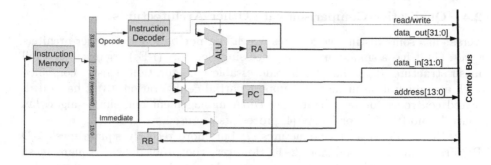

Fig. 6. Controller.

2.3   Controller

Versat uses of a minimal controller for reconfiguration, data transfer and simple algorithmic control. The instruction set contains just 16 instructions for the following actions: (1) loads/stores; (2) basic logic and arithmetic operations; (3) branching. Versat has an accumulator architecture with a 2-stage pipeline shown in Fig. 6. The controller architecture contains 3 main registers: the program counter (PC), the accumulator (RA) and the address register (RB), which is used in indirect loads and stores. There is only one instruction type as illustrated in the figure. The controller is the master of a simple bus called the Control Bus, whose signals are also explained in the figure.

The instruction set is outlined in Table 1. Brackets are used to represent memory pointers. For example, (Imm) represents the contents of the memory position whose address is Imm.

Table 1. Instruction set.

Mnemonic	Opcode	Description
nop	0x0	No operation; PC = PC+1
rdw	0x1	RA = (Imm); PC = PC+1
wrw	0x2	(Imm) = RA; PC = PC+1
rdwb	0x3	RA = (RB); PC = PC+1
wrwb	0x4	(RB) = RA; PC = PC+1
beqi	0x5	RA == 0? PC = Imm: PC = PC+1; RA = RA−1
beq	0x6	RA == 0? PC = (Imm): PC = PC+1; RA = RA−1
bneqi	0x7	RA != 0? PC = Imm: PC = PC+1; RA = RA−1
bneq	0x8	RA != 0? PC = (Imm): PC = PC+1; RA = RA−1
ldi	0x9	RA = Imm; PC = PC+1
ldih	0xA	RA[31:16] = Imm; PC = PC+1
shft	0xB	RA = Imm < 0? RA = RA << 1: RA=RA >> 1
add	0xC	RA = RA+(Imm); PC = PC+1
addi	0xD	RA = RA+Imm; PC = PC+1
sub	0xE	RA = RA−(Imm); PC = PC+1
and	0xF	RA = RA&(Imm); PC = PC+1

2.4   Qualitative Comparison with Other Architectures

Versat has some distinctive features which can not be found in other architectures: (1) it has a small number of processing elements (PEs) organized in a full mesh structure; (2) it has a fully addressable configuration register combined with a configuration memory to support partial configuration; (3) it has a dedicated controller for reconfiguration, DMA management and simple algorithm control – no RISC [3] or VLIW [4] processors are used.

CGRAs started as 1-D structures [7] but more recently square mesh 2-D PE arrays are more common [2–4]. However, the problem with square mesh topologies is that many PEs end up being used as routing resources, reducing the number of PEs available for computation and requiring sophisticated mapping algorithms [10]. Thus, we decided to use a rich interconnect structure and fewer PEs. As explained before, for a small number of PEs, the silicon area occupied by the full mesh interconnect is less than 5% and the limits placed in the frequency of operation are not as stringent as the ones imposed by the energy budgets of certain applications.

It is also important to keep the configuration time to a minimum. As explained in [2], the reconfiguration time in CGRAs can easily dominate the total execution time. To counter this effect we have decided to take partial reconfiguration to the extreme of using a fully addressable configuration register. This keeps the reconfiguration time to a minimum and contrasts with the more moderate hierarchical reconfiguration scheme proposed in [2].

Since it is crucial to have the reconfigurations done quickly, we have decided to include a small dedicated controller with just 16 instructions and low IO latency. It turned out that this controller also proved useful in managing data transfers and running the algorithms of interesting kernels such as the FFT kernel. In other architectures [2–4], more comprehensive processors are used. Our approach reduces the silicon area and power consumption of the core but also limits the complexity of the algorithms that can be run on it. Thus, we rely on other processors that exist in the system to run more complex algorithms, and we restrict Versat to be a kernel accelerator only.

3   Programming

The Versat controller can be programmed using a small C/C++ subset using the Versat compiler. Certain language constructs are interpreted as DE configurations and the compiler automatically generates instructions that write these configurations to the CM. The Versat controller can also be programmed in assembly language, given its easy to apprehend structure. To the best of our knowledge, Versat is the only CGRA that can be programmed in assembly. Despite its simplicity, the Versat controller is able to execute rather complex kernels autonomously.

The purpose of this paper is to describe the Versat architecture, not the Versat programming tools. However, after describing the Controller, it is useful to show an example program to illustrate the features of the C++ subset that

```
int main(){
    //initiate data transfer into Versat using DMA
    dma.config(1024, 256, 256, 1);
    dma.run();

    //clear configuration register and create new configuration
    de.clearConfig();
    for(j = 0; j < 128; j++)
        mem1A[j] = mem0A[j*2] + mem0B[1+j*2];

    //wait for data transfer to finish
    dma.wait();

    //run the data engine
    de.run();

    //configure DMA to transfer result back to memory
    dma.config(2048, 2048+128, 256, 2);

    //wait for data engine to finish
    de.wait(mem1A);

    //transfer result back to memory using DMA
    dma.run();
    dma.wait();
}
```

Fig. 7. Example code.

can be used. The chosen example is the interleaved vector addition program shown in Fig. 7. Comments are added to help understand the code.

Note that the Versat C/C++ dialect does not yet support object or variable declarations. All objects and variables that can be used are predefined. *For* loops with expressions involving memory ports are interpreted as DE configurations and trigger partial reconfigurations. Two nested loops and multiple expressions in the loop body are also supported. Whenever possible, the compiler keeps track of the current state of the configuration register to perform a minimal number of partial reconfigurations needed to prepare the next configuration. Partial reconfigurations generate store instructions in the assembly code, one per configuration field. In many practical situations, the configurations are generated in a program loop where, in each iteration, only a few configuration fields change. A configuration generation loop has a much smaller code footprint than storing all the configurations in memory.

4   Results

Versat has been designed using a UMC 130 nm process. Table 2 compares Versat with a state-of the-art embedded processor and two other CGRA implementations. The Versat frequency and power results have been obtained using the Cadence IC design tools. The power figures have been obtained using the node activity rate extracted from simulating an FFT kernel.

Table 2. Integrated circuit implementation results.

Core	Node (nm)	Area (mm^2)	RAM (kB)	Freq. (MHz)	Power (mW)
ARM Cortex A9 [11]	40	4.6	65.54	800	500
Morphosys [3]	350	168	6.14	100	7000
ADRES [4]	90	4	65.54	300	91
Versat	130	4.2	46.34	170	99

Because the different designs use different technology nodes, to compare the results in Table 2, we need to use a scaling method [12]. A standard scaling method is to assume that the area scales with the square of the feature size and that the power density (W/m^2) remains constant at constant frequency. Doing that, we conclude that Versat is the smallest and least power hungry of the CGRAs. If Versat were implemented in the 40 nm technology, it would occupy about 0.4 mm^2, and consume about 44 mW running at a frequency of 800 MHz. That is, Versat is 10× smaller and consumes about 11× lower power compared with the ARM processor.

The ADRES architecture is about twice the size of Versat. Morphosys is the biggest one, occupying half the size of the ARM processor. These differences can be explained by the different capabilities of these cores. While Versat has a 16-instruction controller and 11 FUs (excluding the memory units), ADRES has a VLIW processor and a 4x4 FU array, and Morphosys has a RISC processor and an 8x8 FU array.

A prototype has been built using a Xilinx Zynq 7010 FPGA, which features a dual-core embedded ARM Cortex A9 system. Versat is connected as a peripheral of the ARM cores using its AXI4 slave interface. The ARM and Versat cores are connected to an on-chip memory controller using their AXI master interfaces. The memory controller is connected to an off-chip DDR module. This FPGA prototype has only been used to measure the execution time in clock cycles. The performance and energy estimates discussed in the next paragraph have been obtained using the measured execution times combined with frequency and power estimates extrapolated from the results in Table 2.

Results for a set of kernels are summarized in Table 3. For both ARM and Versat, the program has been placed in on-chip memory and the data in an external DDR memory device. The Versat *Total* cycle counts include data transfer, processing, control and reconfiguration. The Versat *Unhidden* cycle counts means the control and reconfiguration cycles that do not occur in parallel with the DE or DMA. The average number of FUs used and the code size are given for each kernel. The speedup and energy ratio have been obtained assuming the ARM is running at 800 MHz and Versat is running at 600 MHz in the 40 nm technology. The energy ratio is the ratio between the energy spent by the ARM processor alone and the energy spent by an ARM/Versat combined system using the power figures in Table 2.

Table 3. Kernel benchmark results.

Kernel	ARM Cortex A9 cycles	Versat cycles		Versat FUs used	Versat code size (bytes)	Speedup	Energy ratio
		Total	Unhidden				
vec_add	14726	4517	36	3	152	2.45	2.29
iir1	18890	7487	26	5	220	1.89	1.77
iir2	24488	10567	26	8	332	1.74	1.62
cip	25024	6673	26	10	408	2.81	2.63
fft	394334	16705	624	8.5	3028	17.70	16.55

In Table 3, vec_add is a vector addition, iir1 and iir2 are 1st and 2nd order IIR filters, cip is a complex vector inner product and fft is a Fast Fourier Transform. This kernel set occupies only 50% of the 8 kB program memory. All kernels operate on Q1.31 fixed-point data with vector sizes of 1024. The first 4 kernels use a single Versat configuration and the data transfer size dominates. For example, the vec_add kernel processing time is only 1090 cycles and the remaining 3427 cycles account for data transfer and control. The FFT kernel is more complex and goes through 43 Versat configurations generated on the fly by the Versat controller. The processing time is 12115 cycles and the remaining 4590 cycles is for data transfer and control. It should be noted that most of the reconfiguration and control is done while the data engine is running. In fact, only 605 cycles are unhidden reconfiguration and control cycles in the FFT kernel. In general, this is true for all kernels: unhidden reconfiguration and control cycles are about 1–5% of the total time. The number of FUs used is low for simple kernels like vec_add and iir1 (which could have used a smaller array) but is over 50% for more complex kernels. However, the simpler kernels could have been designed in a more parallel fashion, using more FUs. For example, vec_add can use multiple adders in parallel but only one has been instantiated. These results show good performance speedups and energy savings, even for single configuration kernels.

Most of the 43 FFT configurations derive from two basic configurations that are partially changed many times. The two configurations implement a radix-2 FFT butterfly: one configuration performs the complex product and the other configuration performs the two complex additions in a butterfly. The variations of these configurations alternate the source and destination memories for the data (ping-pong processing), and the address generation constraints to read and write the data values for the various stages and blocks of the FFT. On average only 26% of the array is reconfigured each time. When the array is being configured from the default values, on average 68% of the bits need be written. There is also one configuration to copy the FFT coefficients between two memories (so that they can be accessed from 4 memory ports simultaneously) and to reorder the data by bit reversing the addresses. Partial reconfiguration plays a key role in the FFT example: with full reconfiguration, the execution time grows about 7%, as there are many short loops that are not long enough to overlap with reconfiguration. If full reconfiguration, with all configurations produced at compile time, were used, like in [2–4], the FFT code would be 2.2× larger, penalizing memory bandwidth and capacity.

We can compare Versat with Morphosys since it is reported in [13] that the processing time for a 1024-point FFT is 2613 cycles. Compared with the 12115 cycles taken by Versat, this means that Morphosys was 4.6× faster. This is not surprising since Morphosys has 64 FUs compared to 11 FUs in Versat. However, our point is that an increased area and power consumption is not justified when the CGRA is integrated in a real system. Note that, if scaled to the same technology, Morphosys would be 5× the size of Versat. Unfortunately, comparisons with the ADRES architecture have not been possible, since we have not found any cycle counts published, despite ADRES being one of the most cited CGRA architectures.

It would be nice if we implemented the other approaches in our setup to have a fairer comparison, instead of using published results. However, those are complex cores and implementing them ourselves, besides representing a formidable effort, would carry the risk of us missing some important details that could distort the results. It would also be nice to study Versat's performance in a real application, where it has to be reconfigured periodically to different kernels. We leave that for a more mature state of our evaluation.

5   Conclusion

In this paper we have presented Versat, a minimal CGRA with 4 embedded memories and 11 FUs, a fine partial reconfiguration scheme and a basic 16-instruction controller.

Versat has fewer processing elements compared to other CGRAs (eg. RPU [2]) but uses a full mesh interconnect topology. Another unique feature is a fully addressable configuration register combined with a configuration memory to support partial configuration. The simple Versat controller is used for reconfiguration, DMA management and simple algorithm control, dispensing with complex RISC or VLIW processors proposed in other approaches. The controller generates Versat configurations on the fly, instead of using pre-compiled configurations like other CGRAs. This saves configuration storage space and memory bandwidth. Versat can be programmed in a C/C++ dialect and can be used by host processors by means of an API containing a set of useful kernels.

Results on a VLSI implementation show that Versat is competitive in terms of silicon area (2× smaller than ADRES [4]), and energy consumption (3.6× lower compared with Morphosys [3]). Performance results show that a system combining a state-of-the-art embedded processor and the Versat core can be 17× faster and more energy efficient than the embedded processor alone when running the FFT algorithm.

Acknowledgment. This work was supported by national funds through Fundação para a Ciência e a Tecnologia (FCT) with reference UID/CEC/50021/2013.

References

1. De Sutter, B., Raghavan, P., Lambrechts, A.: Coarse-grained reconfigurable array architectures. In: Bhattacharyya, S.S., Deprettere, E.F., Leupers, R., Takala, J. (eds.) Handbook of Signal Processing Systems, pp. 449–484. Springer, Heidelberg (2010)
2. Liu, L., Wang, D., Zhu, M., Wang, Y., Yin, S., Cao, P., Yang, J., Wei, S.: An energy-efficient coarse-grained reconfigurable processing unit for multiple-standard video decoding. IEEE Trans. Multimed. **17**(10), 1706–1720 (2015)
3. Lee, M.H., Singh, H., Lu, G., Bagherzadeh, N., Kurdahi, F.J.: Design and implementation of the MorphoSys reconfigurable computing processor. J. VLSI Signal Process. Syst. Signal Image Video Technol. **24**, 147–164. Kluwer Academic Publishers (2000)
4. Mei, B., Lambrechts, A., Mignolet, J.-Y., Verkest, D., Lauwereins, R.: Architecture exploration for a reconfigurable architecture template. Des. Test Comput. **22**(2), 90–101 (2005)
5. de Sousa, J.T., Martins, V.M.G., Lourenco, N.C.C., Santos, A.M.D., do Rosario Ribeiro, N.G.: Reconfigurable coprocessor architecture template for nested loops and programming tool. US Patent 8,276,120 (2012)
6. Hartenstein, R., Herz, M., Hoffmann, T., Nageldinger, U.: Mapping applications onto reconfigurable KressArrays. In: Lysaght, P., Irvine, J., Hartenstein, R. (eds.) FPL 1999. LNCS, vol. 1673, pp. 385–390. Springer, Heidelberg (1999). doi:10.1007/978-3-540-48302-1_42
7. Ebeling, C., Cronquist, D.C., Franklin, P.: RaPiD — reconfigurable pipelined datapath. In: Hartenstein, R.W., Glesner, M. (eds.) FPL 1996. LNCS, vol. 1142, pp. 126–135. Springer, Heidelberg (1996). doi:10.1007/3-540-61730-2_13
8. Baumgarte, V., Ehlers, G., May, F., Nückel, A., Vorbach, M., Weinhardt, M.: PACT XPP - a self-reconfigurable data processing architecture. J. Supercomput. **26**(2), 167–184 (2003)
9. Farahini, N., Hemani, A., Sohofi, H., Jafri, S.M.A.H., Tajammul, M.A., Paul, K.: Parallel distributed scalable runtime address generation scheme for a coarse grain reconfigurable computation and storage fabric. Microprocess. Microsyst. **38**(8), 788–802 (2014)
10. Liu, D., Yin, S., Liu, L., Wei, S.: Polyhedral model based mapping optimization of loop nests for CGRAs. In: 2013 50th ACM/EDAC/IEEE Design Automation Conference (DAC), pp. 1–8 (2013)
11. Wang, W., Dey, T.: A survey on ARM Cortex A processors. http://www.cs.virginia.edu/skadron/cs8535s11/armcortex.pdf. Accessed 6 Apr 2016
12. Huang, W., Rajamani, K., Stan, M.R., Skadron, K.: Scaling with design constraints: predicting the future of big chips. IEEE Micro **31**(4), 16–29 (2011)
13. Kamalizad, A.H., Pan, C., Bagherzadeh, N.: Fast parallel FFT on a reconfigurable computation platform. In: 15th Symposium on Computer Architecture and High Performance Computing, Proceedings, pp. 254–259 (2003)

Environments/Libraries to Support Parallelization

An Application-Level Solution for the Dynamic Reconfiguration of MPI Applications

Iván Cores[1], Patricia González[1], Emmanuel Jeannot[2], María J. Martín[1(✉)],
and Gabriel Rodríguez[1]

[1] Grupo de Arquitectura de Computadores, Universidade da Coruña,
A Coruña, Spain
mariam@udc.es
[2] INRIA Bordeaux Sud-Ouest, Bordeaux, France

Abstract. Current parallel environments aggregate large numbers of computational resources with a high rate of change in their availability and load conditions. In order to obtain the best performance in this type of infrastructures, parallel applications must be able to adapt to these changing conditions. This paper presents an application-level proposal to automatically and transparently adapt MPI applications to the available resources. The architecture includes: automatic code transformation of the parallel applications, a system to reschedule processes on available nodes, and migration capabilities based on checkpoint-and-restart techniques to move selected processes to target nodes. Experimental results show a good degree of adaptability and a good performance in different availability scenarios.

Keywords: HPC · MPI · Checkpointing · Migration · Scheduling

1 Introduction

The resources availability of large-scale distributed systems may vary during a job execution, making malleable applications, that is, parallel programs that are able to adapt their execution to the number of available processors at runtime, specially appealing. Malleable jobs provide important advantages for the final users and the whole system, like a higher productivity and a better response time [3,9], or a greater resilience to node failures [5]. These characteristics will allow to improve the use of resources, which will have a direct effect on the energy consumption required for the execution of applications, resulting in both cost savings and a greener computing.

High performance computing (HPC) is nowadays dominated by the MPI paradigm. Most MPI applications follow the SPMD (single program, multiple data) programming model and they are executed in HPC systems by specifying a fixed number of processes running on a fixed number of processors. The resource allocation is statically specified during job submission, and maintained constant during the entire execution. Thus, MPI applications are unable to dynamically adapt to changes in resource availability.

© Springer International Publishing AG 2017
I. Dutra et al. (Eds.): VECPAR 2016, LNCS 10150, pp. 191–205, 2017.
DOI: 10.1007/978-3-319-61982-8_18

The aim of this work is to propose a solution to transform existing MPI applications into malleable jobs, that is, jobs that are capable of adapting their executions to changes in the environment. The proposed solution is based on process migration, thus, if a node becomes unavailable, the processes on that node will be migrated to other available ones, overloading nodes when necessary. Checkpoint and restart is used to implement this migration. The state of each process to be migrated is stored into memory and transferred to a new available node. Afterwards, this state is recovered by a newly created process, which continues the execution. To this end, the proposal is implemented at the application-level, extending the functionalities of the CPPC (ComPiler for Portable Checkpointing) framework [18]. CPPC is an application-level checkpointing tool, available under GNU general public license (GPL) at http://cppc.des.udc.es, that appears to the final user as a compiler tool and a runtime library. At compile time, the CPPC source-to-source compiler automatically transforms a parallel code into an equivalent version with calls to the CPPC library to instrument the dynamic reconfiguration.

The structure of this paper is as follows. Section 2 describes related work. Section 3 describes the solution proposed to automatically and transparently transform existing MPI application into malleable jobs. Section 4 evaluates the performance of the proposal. Finally, Sect. 5 concludes the paper.

2   Related Work

There are several proposals in the literature that use a stop-and-restart approach to implement malleable MPI applications [1,16,20]. However, stop and restart solutions imply a job requeueing, with the consequent loss of time. Dynamic reconfiguration [8,12,17,22], on the other hand, allows to change the number of processors while the program is running without having to stop and relaunch the application. Most of these solutions [12,17,22] change the number of processes to adapt to the number of available processors, which implies a redistribution of the data and, thus, they are very restrictive with the kind of application they support. On the other hand, in AMPI [8] the number of processes is preserved and the application adapts to changes in the number of resources through migration based on virtualization.

Besides AMPI, there exist in the literature different research works that provide process migration through the use of virtualization technologies [7,13]. However, virtualization solutions present important performance penalties due to larger memory footprints [21]. Moreover, the performance of MPI applications relies heavily on the performance of communications. Currently virtualization of CPU and memory resources presents a low overhead. However, the efficient virtualization of network I/O is still work in progress. For instance, recent results in migration over Infiniband networks [6] show very high overhead, strong scalability limitations and tedious configuration issues.

3   Reconfiguration of MPI Applications

The aim of this work is to build MPI applications that are able to dynamically adapt their execution to changes in the resource availability. The following subsections describe the main components of the proposed solution: (1) the triggering of the reconfiguration operation; (2) the scheduling algorithm implemented to allow the application to autonomously decide which processes should be moved and to which target nodes; and (3) the migration operation itself. The main steps of the reconfiguration process are depicted in Fig. 1.

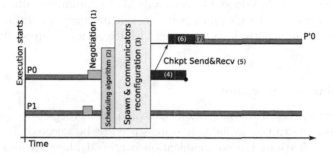

Fig. 1. Steps in the reconfiguration operation: (1) Negotiation to select a single safe location to trigger the reconfiguration; (2) Scheduling algorithm to decide the processes to be migrated and the new allocations; (3) Spawning of new processes and reconfiguration of the communicators; (4) Checkpointing of the migrating processes; (5) Sending of the checkpoint files; (6) Recovering the state from the checkpoint files; and (7) Effective recovering of the application state. Steps (4) to (6) are partially overlapped.

3.1   Triggering the Reconfiguration Operation

The proposed solution is based on dynamic live process migration. If a node becomes unavailable, the processes on that node will be migrated to other available ones (without stopping the application), overloading nodes when necessary. Our proposal relies on a monitoring system that provides dynamical information about the available resources. There are in the literature many proposals for different environments and objectives. For this work we assume that an availability file is set up for each malleable MPI job. This file contains the names of all the nodes that are assigned to execute the MPI job together with their number of available cores. A change in this file activates a flag in the CPPC controller to start the reconfiguration of the execution. During this reconfiguration, some MPI processes will be migrated and, thus, communication groups must be rebuilt. The reconstruction of the communication groups is a critical step, since replacing communicators may lead to an inconsistent global state: messages sent/received using the old communicators cannot be received/sent using the new ones. A possible solution to this problem is to restrict the points at which the migration can be started, making the reconstruction of the communicators, and thus the reconfiguration, in locations where there are no pending communications, i.e., safe points. The CPPC compiler automatically detects safe points, thus facilitating the implementation of this approach. However, conducting the reconfiguration from different safe points in different processes may lead to inconsistencies. Therefore, a negotiation protocol is needed at runtime to select a single safe location as the place to trigger the reconfiguration operation to achieve a consistent global state after migration. This location will be the next safe point to be reached by the process that has advanced the farthest in the execution. Each process communicates to all other processes the last safe point it has crossed. One-sided asynchronous MPI communications are used so that processes may continue running without synchronizations during the negotiation, overlapping negotiation operations with execution progress and avoiding deadlocks.

3.2 Scheduling Algorithm

Previous to the start of the migration operation, the processes to be migrated and their mapping to the available resources need to be selected.

In MPI applications the communication overhead plays an important role in the global performance of the parallel application. To be able to migrate the processes efficiently we need to know the *affinity* between the processes so as to map those with a high communication rate as close to each other as possible. For this aim TreeMatch [11] and Hwloc [2] are used. TreeMatch is an algorithm that obtains the optimized process placement based on the communication pattern among the processes and the hardware topology of the underlying computing system. It tries to minimize the communications at all levels, including network, memory and cache hierarchy. It takes as input both a matrix modeling the communications among the processes, and a representation of the topology of the system. The topological information, represented as a tree, is provided by Hwloc and obtained dynamically during application execution. TreeMatch returns as output an array with the core ID that should be assigned to each process.

An example of the output of TreeMatch is given in Fig. 2. On the left, a communication matrix representing the affinity between processes is given as input: the darker the dot the higher the communication volume and hence the affinity. TreeMatch computes the permutation (σ) of the processes such that the cores with high affinity are mapped close together on the tree representing the target topology. After applying the permutation the communication matrix on the right is obtained.

The communication matrix needed by TreeMatch is obtained dynamically, just before the scheduling algorithm is triggered, using a monitoring component developed for Open MPI and integrated in an MCA (Modular Component Architecture) framework called *pml* (point-to-point management layer). This component, if activated at launch time (through the mpiexec option --mca pml_monitoring_enable), monitors all the communications at the lower level of Open MPI (i.e., once collective communications have been decomposed into

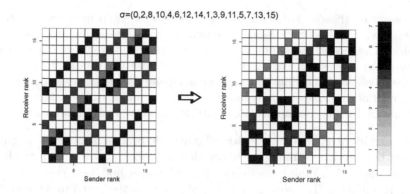

Fig. 2. TreeMatch example for a binary tree of 4 levels and 16 leaves (cores).

send/recv communications). Therefore, contrary to the MPI standard profiling interface (PMPI) approach where the MPI calls are intercepted, here the actual point-to-point communications that are issued by Open MPI are monitored, which is much more precise. This monitoring component was previously developed by one of the authors and it will be released in Open MPI 2.0.

To evaluate the overhead of this monitoring, the execution of the LU NAS benchmark with and without monitoring have been compared. The LU kernel has been selected since it is the one that sends the largest number of messages. Results for this kernel are shown in Table 1. Shown times are the average of 10 runs. Results using 16 and 64 processes using classes A, B and C show that the overhead is very low (less than 0.7%). Moreover, it decreases with the class of the kernel (i.e., the problem size), being the overhead to manage one message 1 µs or less.

TreeMatch focuses on minimizing the communication cost in a parallel execution. Thus, if TreeMatch is directly applied to find the processes mapping during a reconfiguration phase, it could lead to a complete replacement of all the application processes. This would involve unnecessary process migrations and, thus,

Table 1. Monitoring overhead for the LU NAS Kernel on nodes with 2 quad-core Nehalem Intel Xeon processors interconnected by an InfiniBand QDR network.

Class	Number of processes	Total number of messages	Number of mess. per processes	Exec. time (s)	Monitoring exec. time (s)	Overhead
A	16	380630	23789.375	5.696	5.72	0.42%
B	16	609542	38096.375	23.155	23.189	0.15%
C	16	970982	60686.375	90.665	90.727	0.10%
A	64	1777226	27769.156	1.732	1.744	0.69%
B	64	2845482	44460.656	6.522	6.567	0.69%
C	64	4532202	70815.656	25.335	25.374	0.15%

unnecessary overheads. To avoid this behavior, a two-step mapping algorithm was designed. The first step decides the number and the specific processes to be migrated. The second step finds the best target nodes and cores to place these processes. An interesting feature of TreeMatch is that the topology given as an input can be a real machine topology or a virtual topology designed to separate groups of processes in clusters such that communications within clusters are maximized while communications outside the clusters are minimized.

– **Step 1: identify processes to migrate**. A process should be migrated either because it is running on nodes that are going to become unavailable, or because it is running on oversubscribed nodes and new resources have become available. To know the number of processes that need to be migrated, all processes exchange, via MPI communications, the numbers of the node and core in which they are currently running. Then, using this information, each application process calculates the current computational load of each node listed in the availability file associated to the application. A *load* array is computed, where $load(i)$ is the number of processes that are being executed in node n_i. Besides, each process also calculates the maximum number of processes that could be allocated to each node n_i in the new configuration:

$$maxProcs(i) = \left\lceil nCores(i) \times \frac{N}{nTotalCores} \right\rceil$$

where $nCores(i)$ is the number of available cores of node n_i, N is the number of processes of the MPI application, and $nTotalCores$ is the number of total available cores. If $load(i) > maxProcs(i)$ then $load(i) - maxProcs(i)$ processes have to be migrated. If the node is no longer available, $maxProcs(i)$ will be equal to zero and all the processes running in that node will be identified as migrating processes. Otherwise, TreeMatch is used to identify the migrating processes. The aim is to maintain in each node the most related processes according to the application communication pattern. Figure 3 illustrates an example with two 16-core nodes executing a 56-process application in an oversubscribed scenario. When two new nodes become available, 12 processes per node should be migrated to the new resources. To find the migrating processes, TreeMatch is queried once for each oversubscribed node. The input is a virtual topology breaking down the node into two virtual ones: one with maxProcs(i) cores and the other with load(i) − maxProcs(i) cores. TreeMatch uses in runtime this virtual topology, and a sub-matrix with the communication pattern between the processes currently running on the node, to identify the processes to be migrated, that is, those mapped to the second virtual node.

– **Step 2: identify target nodes**. Once the processes to be migrated are identified, the target nodes (and the target cores inside the target nodes) to place these processes have to be found. TreeMatch is again used to find the best placement for each migrating process. It uses a sub-matrix with the communication pattern of the migrating processes, and a virtual topology built from the real topology of the system but restricted to use only the potential target nodes in the cluster. The potential targets are those nodes that satisfy $load(i) < maxProcs(i)$. They can be empty nodes, nodes already in use but with free cores, or nodes that need to be oversubscribed. Since TreeMatch only allows the mapping of one process per core, if there are no sufficient real target cores to allocate the migrating processes, a virtual topology will simulate $maxProcs(i) - load(i)$ extra cores in the nodes that need to be oversubscribed. Figure 4 illustrates the second step of the algorithm for the same example of Fig. 3. In this example, the virtual topology used consists of the new available nodes in the system, two 16-core nodes to map the 24 processes. After executing TreeMatch, the target cores and, therefore, the target nodes for the migrating processes obtained in step 1 are identified and CPPC can be used to perform the migration.

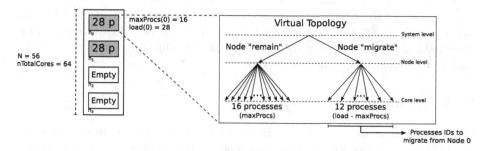

Fig. 3. Step 1: identifying processes to be migrated. Virtual topology built to migrate 12 processes from a 16-core node where 28 processes are running (16 processes remain and 12 processes migrate).

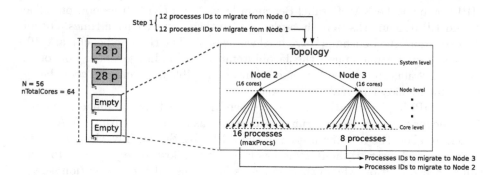

Fig. 4. Step 2: identifying target nodes. Topology built to map the migrating processes selected in step 1 to the empty cores in the system.

3.3 Migration Operation

Once the mapping of migrated processes to available resources is decided, the migration operation can start. First, new processes are spawned in the target nodes to replace the migrating ones, and the global communicator is reconstructed. The dynamic process management facilities of MPI-2 are used for these operations. Then, the migrating processes save their state, storing it into memory. The checkpoint files of the terminating processes are sent using MPI communications. At this point the terminating processes can safely finalize. Then, the new processes restart the execution by reading the checkpoint files and recovering the application state. This is achieved by delegating to CPPC and employing its native capabilities. The procedure is depicted in Fig. 1.

Initially the new spawned processes are not bound to any specific core. The TreeMatch assignment is sent to the new processes together with the checkpoint file and CPPC performs the binding via the Hwloc library.

To minimize the overhead associated to the I/O operations needed for the migration, the checkpoint files are split into several chunks and transferred in a pipelined fashion, overlapping the writing in the terminating processes with the reading in the newly spawned ones [19].

4 Experimental Results

This section aims to show the feasibility of the proposal and to evaluate the cost of the reconfiguration whenever a change in the resource availability occurs. A multicore cluster was used to carry out these experiments. It consists of 16 nodes, each powered by two octa-core Intel Xeon E5-2660 CPUs with 64 GB of RAM. The cluster nodes are connected through an InfiniBand FDR network. The working directory is mounted via network file system (NFS) and it is connected to the cluster by a Gigabit Ethernet network.

The application testbed is composed of six out of the eight applications in the MPI version of the NAS Parallel Benchmarks v3.1 [14] (NPB from now on). The IS and EP benchmarks were discarded due to their low execution times. For all the executions the benchmark size used was class C. The Himeno benchmark [10] was also tested. Himeno uses the Jacobi iteration method to solve the Poissons's equation, evaluating the performance of incompressible fluid analysis code, being a benchmark closer to real applications.

The MPI implementation used was Open MPI v1.8.1 [15] modified to enable dynamic monitoring. The `mpirun` environment has been tuned using MCA parameters to allow the reconfiguration of the MPI jobs. Specifically, the parameter `orte_allowed_exit_without_sync` has been set to allow some processes to continue their execution when other processes have finished their execution safely. The parameter `mpi_yield_when_idle` was also set to force degraded execution mode and, thus, to allow progress in an oversubscribed scenario.

To evaluate the feasibility of the proposed solution and its performance, different scenarios have been forced during the execution of the applications. Figure 5 illustrates these scenarios. The applications were initially launched in a 64-process configuration running on 4 available nodes of the cluster (16 cores per node). Then, after a time, one of the nodes becomes unavailable. In this scenario, the 16 processes running on the first node should be moved to the empty node, and the application execution continues in a 4 node configuration. After a while, the 4 nodes where the application is running start to become unavailable sequentially, first one, then another, without spare available nodes to replace them, until only one node is available and the 64 processes are running on it. Finally, in a single step, the last node fails but 4 nodes become available again, and the processes are migrated to return again to using 4 nodes. To demonstrate the feasibility of the solution, the iteration time was measured across the execution in those scenarios. Measuring iteration time allows to have a global vision of the instantaneous performance impact.

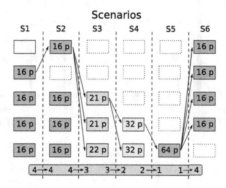

Fig. 5. Selected scenarios to show the feasibility of the proposal and evaluate the migration and scheduling cost.

Figure 6 shows the results for all the benchmarks in the scenarios illustrated in Fig. 5. These results demonstrate that, using the proposed solution, the applications are capable of adjusting their execution to changes in the environment. The high peaks in these figures correspond to reconfiguration points. For comparison purposes, the compute times that would be attained if the application could adjust its granularity to the available resources[1], instead of oversubscribing them without modifying the original number of processes are also shown (green line). The overhead that would introduce the data distribution needed to adjust the application granularity is not shown in the figure.

Table 2 details the main impacting steps in the reconfiguration overhead for all the NPB applications. As shown in Fig. 1, the iteration time when a reconfiguration is performed can be broken down into the following stages:

- *Negotiation*: execution time of the negotiation protocol used to reach consensus on the reconfiguration point.

---

[1] This time is measured executing the application with a different number of processes depending on the hardware available (16 processes version when only 1 node is available, 32 processes version when 2 nodes are available, etc.).

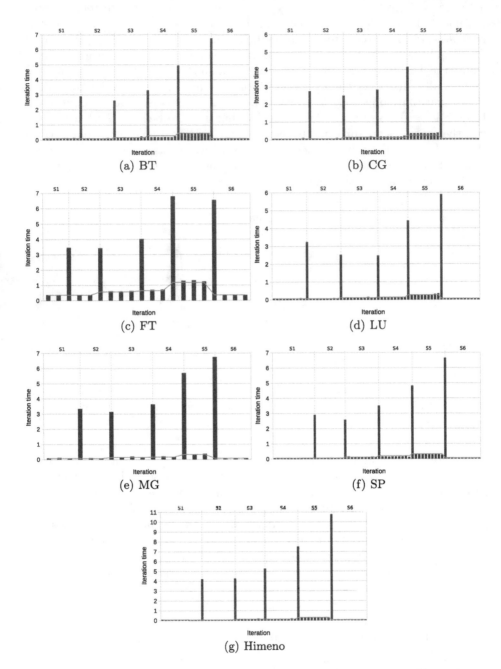

Fig. 6. Iteration execution times in the scenarios illustrated in Fig. 5. (Color figure online)

- *Scheduling*: execution time of the scheduling algorithm to identify processes to be moved and target nodes.
- *Spawn&Rec*: execution time of the spawn function and the reconfiguration of the communicators.
- *ChkptTransfer&Read*: average time to write the checkpoint files in the terminating processes, transfer them to target nodes, and read them in newly spawned processes.
- *Restart*: average time to complete the restart of the application once the checkpoint files have been read.
- *Compute*: the computational time of the iteration where the reconfiguration takes place.

Table 2. Execution time (s) of the reconfiguration phases.

	scenarios	NPB applications						Himeno
		BT	CG	FT	LU	MG	SP	
Negotiation	1 → 2	0.88	1.18	0.01	1.08	0.93	0.94	1.12
	2 → 3	0.80	0.85	0.01	0.86	0.94	0.81	0.90
	3 → 4	0.88	0.89	0.14	0.53	0.94	0.86	0.95
	4 → 5	1.06	1.11	0.02	1.15	1.07	1.04	1.09
	5 → 6	1.86	1.94	0.02	1.90	1.88	1.87	1.95
Spawn&Rec	1 → 2	1.32	1.18	1.44	1.61	1.28	1.29	1.13
	2 → 3	1.15	1.18	1.11	1.08	1.09	1.14	0.99
	3 → 4	1.43	1.43	1.32	1.31	1.35	1.59	1.48
	4 → 5	2.23	2.14	2.14	2.17	2.38	2.07	2.12
	5 → 6	3.27	3.19	3.25	3.28	3.20	3.22	3.23
ChkptTransfer&Read	1 → 2	0.35	0.10	0.39	0.16	0.44	0.39	1.44
	2 → 3	0.37	0.11	0.43	0.18	0.42	0.40	1.59
	3 → 4	0.48	0.14	0.47	0.21	0.54	0.54	1.96
	4 → 5	0.75	0.23	0.84	0.36	0.84	0.93	2.92
	5 → 6	1.01	0.28	1.41	0.47	1.41	1.06	4.96
Restart	1 → 2	0.15	0.01	0.04	0.01	0.04	0.12	0.25
	2 → 3	0.05	0.01	0.20	0.02	0.03	0.08	0.46
	3 → 4	0.24	0.01	0.26	0.02	0.03	0.30	0.47
	4 → 5	0.36	0.02	0.41	0.02	0.06	0.30	0.57
	5 → 6	0.36	0.01	0.41	0.01	0.03	0.29	0.45

The *Negotiation* phase depends on the application as in this phase MPI one-sided communications are used and the progress of these remote operations is affected by the MPI calls inside the application. These times could be lower using other MPI implementations and/or computer architectures [4].

The largest contribution to the reconfiguration cost is due to the *Spawn&Rec* step. The time spent in the spawn function depends on the number of spawned processes and the degree of oversubscription. The more processes to be migrated, the larger the overhead of this phase. This can be observed comparing the overhead associated to the reconfiguration from scenario 1 to scenario 2, where 16 processes are moved to an empty target node, and the overhead associated to the reconfiguration from scenario 5 to scenario 6, where 64 processes are moved to 4 empty target nodes. When target nodes are oversubscribed, the computation time of each process is penalized and so is the *Spawn&Rec* phase, specially affected due to their collective communications. This can be observed in the increase that the overhead of the *Spawn&Rec* phase suffers in the reconfiguration from scenario 2 to scenario 3, from scenario 3 to scenario 4, from scenario 4 to scenario 5, and from scenario 5 to scenario 6, where 16, 21, 32 and 64 processes are migrated each time, oversubscribing the surviving nodes. Finally, since this phase involves different collective communications, its time depends on the total number of processes in the application. This can be observed in Table 3, that shows the overhead of the *Spawn&Rec* step when migrating 16 processes to an empty target node with a different number of processes in the application.

Table 3. Overhead (in seconds) of the *Spawn&Rec* step when spawning 16 new processes vs total number of processes in the application.

NPB	Number of total processes			
	16	32/36	64	128/121
BT	0.97	0.99	1.32	1.57
CG	0.98	1.01	1.18	1.79
FT	0.96	1.07	1.44	1.75
LU	1.00	1.01	1.61	1.89
MG	1.02	1.01	1.28	1.63
SP	0.99	1.00	1.29	1.72
Himeno	0.99	1.01	1.13	1.96

The *ChkptTransfer&Read* step also impacts significantly the reconfiguration overhead. The I/O operations are recognized to be one of the main impacting factors in the performance of migration operations, specially in those associated to checkpoint solutions. Checkpoint file sizes are critical to minimize the I/O time. CPPC applies live variable analysis and identification of zero-blocks to decrease checkpoint file sizes. Table 4 shows the checkpoint sizes per process and the total data size transferred between nodes when migrating 16, 32 and 64 processes. The total amount of data varies between 127 MB for CG migrating a single node (16 processes) and 10.42 GB for Himeno when migrating 4 nodes (64 processes). By means of a pipelined approach [19] that overlaps the state file writing in the terminating processes, the data transfer through the network,

Table 4. Transfer size (checkpoint size in MB).

NPB	Checkpoint size per process	Total data size migrated		
		16 proc.	32 proc.	64 proc.
BT	33.15	530.45	1060.90	2121.80
CG	7.93	126.97	253.95	507.90
FT	48.09	769.50	1539.01	3078.02
LU	15.48	247.74	495.48	918.96
MG	39.26	628.19	1256.39	2512.78
SP	32.12	513.99	1027.99	2055.98
Himeno	166.71	2667.36	5334.72	10669.44

and the state file read in the new processes, the proposed solution achieves to significantly reduce this impact.

The *Restart* step is a small contributor to the reconfiguration overhead. An important part of this time is derived from the negotiation protocol used. During the negotiation phase each process specifies a memory region (window) that it exposes to others. Since the MPI communicators of the application have been reconfigured, at restart time the old MPI windows have to be closed and new ones have to be created. Although not as impacting as the spawning function, the overhead of this operation is not negligible.

Finally, the *Scheduling* phase is negligible for all the NPB applications, being always smaller than 0.1 s. Thus, these times are not included in the table.

5   Concluding Remarks

In this paper a proposal to automatically and transparently adapt MPI applications to available resources is proposed. The solution relies on a previous application-level migration approach, incorporating a new scheduling algorithm based on TreeMatch, Hwloc and dynamic communication monitoring, to obtain well balanced nodes while preserving performance as much as possible.

The experimental evaluation of the proposal shows successful and efficient operation, with an overhead of a few seconds during reconfiguration, which will be negligible in large applications with a more realistic reconfiguration frequency.

Proposals like the one in this paper will be of particular interest in future large scale computing systems, since applications that are able to dynamically reconfigure themselves to adapt to different resource scenarios will be key to achieve a tradeoff between energy consumption and performance.

Acknowledgments. This research was partially supported by the Ministry of Economy and Competitiveness of Spain and FEDER funds of the EU (Project TIN2013-42148-P), by the Galician Government and FEDER funds of the EU (consolidation program of competitive reference groups GRC2013/055) and by the EU under the COST programme Action IC1305, Network for Sustainable Ultrascale Computing.

References

1. Agbaria, A., Friedman, R.: Starfish: fault-tolerant dynamic MPI programs on clusters of workstations. Cluster Comput. **6**(3), 227–236 (2003)
2. Broquedis, F., Clet-Ortega, J., Moreaud, S., Furmento, N., Goglin, B., Mercier, G., Thibault, S., Namyst, R.: hwloc: a generic framework for managing hardware affinities in HPC applications. In: Euromicro International Conference on Parallel, Distributed and Network-Based Processing (PDP), pp. 180–186 (2010)
3. Buisson, J., Sonmez, O., Mohamed, H., Lammers, W., Epema, D.: Scheduling malleable applications in multicluster systems. In: 2007 International Conference on Cluster Computing (CLUSTER), pp. 372–381 (2007)
4. Cores, I., Rodríguez, G., Martín, M.J., González, P.: Achieving checkpointing global consistency through a hybrid compile time and runtime protocol. Procedia Comput. Sci. **18**, 169–178 (2013). International Conference on Computational Science (ICCS)
5. George, C., Vadhiyar, S.S.: ADFT: an adaptive framework for fault tolerance on large scale systems using application malleability. Procedia Comput. Sci. **9**, 166–175 (2012). International Conference on Computational Science (ICCS)
6. Guay, W.L., Reinemo, S.A., Johnsen, B.D., Yen, C.H., Skeie, T., Lysne, O., Tørudbakken, O.: Early experiences with live migration of SR-IOV enabled infiniband. J. Parallel Distrib. Comput. **78**, 39–52 (2015)
7. Hacker, T.J., Romero, F., Nielsen, J.J.: Secure live migration of parallel applications using container-based virtual machines. Int. J. Space Based Situated Comput. **2**(1), 45–57 (2012)
8. Huang, C., Lawlor, O., Kalé, L.V.: Adaptive MPI. In: Rauchwerger, L. (ed.) LCPC 2003. LNCS, vol. 2958, pp. 306–322. Springer, Heidelberg (2004). doi:10.1007/978-3-540-24644-2_20
9. Hungershofer, J.: On the combined scheduling of malleable and rigid jobs. In: Computer Architecture and High Performance Computing (SBAC-PAD), pp. 206–213 (2004)
10. Information Technology Center, RIKEN. HIMENO Benchmark. http://accc.riken.jp/2444.htm. Accessed Aug 2016
11. Jeannot, E., Mercier, G.: Near-optimal placement of MPI processes on hierarchical NUMA architectures. In: D'Ambra, P., Guarracino, M., Talia, D. (eds.) Euro-Par 2010. LNCS, vol. 6272, pp. 199–210. Springer, Heidelberg (2010). doi:10.1007/978-3-642-15291-7_20
12. Martín, G., Singh, D.E., Marinescu, M.C., Carretero, J.: Enhancing the performance of malleable MPI applications by using performance-aware dynamic reconfiguration. Parallel Comput. **46**, 60–77 (2015)
13. Nagarajan, A.B., Mueller, F., Engelmann, C., Scott, S.L.: Proactive fault tolerance for HPC with Xen virtualization. In: International Conference on Supercomputing (ICS), pp. 23–32 (2007)
14. National Aeronautics and Space Administration. The NAS Parallel Benchmarks. http://www.nas.nasa.gov/publications/npb.html. Accessed Aug 2016
15. Open MPI Team. Open MPI: Open Source High Performance Computing. http://www.open-mpi.org/. Accessed Aug 2016
16. Raveendran, A., Bicer, T., Agrawal, G.: A framework for elastic execution of existing MPI programs. In: IEEE International Symposium on Parallel and Distributed Processing Workshops (IPDPSW), pp. 940–947 (2011)

17. Ribeiro, F.S., Nascimento, A.P., Boeres, C., Rebello, V.E.F., Sena, A.C.: Autonomic malleability in iterative MPI applications. In: Computer Architecture and High Performance Computing (SBAC-PAD), pp. 192–199 (2013)
18. Rodríguez, G., Martín, M.J., González, P., Touriño, J., Doallo, R.: CPPC: a compiler-assisted tool for portable checkpointing of message-passing applications. Concurr. Comput. Pract. Exper. **22**(6), 749–766 (2010)
19. Rodríguez, M., Cores, I., González, P., Martín, M.J.: Improving an MPI application-level migration approach through checkpoint file splitting. In: Computer Architecture and High Performance Computing (SBAC-PAD), pp. 33–40 (2014)
20. Vadhiyar, S.S., Dongarra, J.J.: SRS - a framework for developing malleable and migratable parallel applications for distributed systems. Parallel Process. Lett. **13**(02), 291–312 (2003)
21. Wang, C., Mueller, F., Engelmann, C., Scott, S.L.: Proactive process-level live migration and back migration in HPC environments. J. Parallel Distrib. Comput. **72**(2), 254–267 (2012)
22. Weatherly, D.B., Lowenthal, D.K., Nakazawa, M., Lowenthal, F.: Dyn-MPI: supporting MPI on non dedicated clusters. In: ACM/IEEE Conference on High Performance Networking and Computing (SC), p. 5 (2003)

Scientific Workflow Scheduling with Provenance Support in Multisite Cloud

Ji Liu[1]([✉]), Esther Pacitti[1], Patrick Valduriez[1], and Marta Mattoso[2]

[1] Inria, Microsoft-Inria Joint Centre, LIRMM and University of Montpellier,
Montpellier, France
{ji.liu,patrick.valduriez}@inria.fr, esther.pacitti@lirmm.fr
[2] COPPE, Federal University of Rio de Janeiro, Rio de Janeiro, Brazil
marta@cos.ufrj.br

Abstract. Recently, some Scientific Workflow Management Systems (SWfMSs) with provenance support (*e.g.* Chiron) have been deployed in the cloud. However, they typically use a single cloud site. In this paper, we consider a multisite cloud, where the data and computing resources are distributed at different sites (possibly in different regions). Based on a multisite architecture of SWfMS, *i.e.* multisite Chiron, we propose a multisite task scheduling algorithm that considers the time to generate provenance data. We performed an extensive experimental evaluation of our algorithm using Microsoft Azure multisite cloud and two real-life scientific workflows (Buzz and Montage). The results show that our scheduling algorithm is up to 49,6% better than baseline algorithms in terms of total execution time.

Keywords: Scientific workflow · Scientific workflow management system · Scheduling · Parallel execution · Multisite cloud

1 Introduction

Many large-scale *in silico* scientific experiments take advantage of scientific workflows (SWfs) to model data operations such as loading input data, data processing, data analysis and aggregating output data. SWfs enable scientists to model the data processing of these experiments as a graph, in which vertices represent data processing activities and edges represent dependencies between them. A SWf is the assembly of scientific data processing activities with data dependencies between them [5]. An activity is a description of a piece of work that forms a logical step within a SWf representation [12] and a task is the representation of an activity within a one-time execution of the activity. Since the tasks of the same activity process different data chunks [12], they are independent. We assume that the tasks have similar workloads in this paper. A Scientific Workflow Management System (SWfMS) is a tool to execute SWfs [12]. Some implementations of SWfMSs are publicly available, *e.g.* Pegasus [7] and Chiron [15]. A SWfMS generally supports provenance data, which is the metadata

© Springer International Publishing AG 2017
I. Dutra et al. (Eds.): VECPAR 2016, LNCS 10150, pp. 206–219, 2017.
DOI: 10.1007/978-3-319-61982-8_19

that captures the derivation history of a dataset [12], during SWf execution. In order to execute a data-intensive SWf within a reasonable time, SWfMSs generally exploit High Performance Computing (HPC) resources obtained from a computer cluster, grid or cloud environment.

Recently, some SWfMSs with provenance support (*e.g.* Chiron) have been proposed for a single cloud site. However, the input data necessary to run a SWf may well be distributed at different sites (possibly in different regions) and may not be allowed to be transferred to other sites, *e.g.* because of big size or proprietary reasons. Furthermore, it may not be always possible to move all the computing resources (including programs) to a single site. In this paper, we consider a multisite cloud composed of several sites (or data centers) of the same cloud provider, each with its own resources and data. The difference between multisite cloud and the environment of single-site or supercomputer is that, in multisite cloud, the data or the computing resources are distributed at different sites and the network bandwidths among different sites are different. Compared with P2P, a major difference is that multisite cloud does not have many sites.

To enable SWf execution in a multisite cloud with distributed input data, the execution of the tasks of each activity should be scheduled to cloud sites (or sites for short). The tasks of each activity can be scheduled independently. Then, the scheduling problem is how to decide at which sites to execute the tasks of each activity in order to reduce execution time of a SWf in a multisite cloud. The mapping relationship between sites and tasks is a scheduling plan. Since it may take a significant amount of time to transfer data between two different sites, the multisite scheduling problem should take into account the resources at different sites and intersite data transfer, including intermediate data to be processed by tasks and the provenance data.

Classic scheduling algorithms, *e.g.* Opportunistic Load Balancing (OLB) [14], Minimum Completion Time (MCT) [14], min-min [10], max-min [10] and Heterogeneous Earliest Finish Time (HEFT) [19], and some others [4,17,18] are designed to address the scheduling problem within a single site. Although some may be used for multiple sites, they do not provide support for provenance data. A few multisite scheduling approaches are proposed [9], but they do not consider the distribution of input data at different sites and have no support for provenance data, which may incur much time for intersite data transfer. In [16], data transfer is analyzed in multisite SWf execution, stressing the importance of optimizing data provisioning. However, this information is not yet explored on task scheduling. In our previous works [11,13], we proposed the solutions of scheduling the execution of each activity to a site, which cannot schedule the tasks of one activity to different sites to process the distributed data.

The difference between our work and others is multisite execution with provenance support. In the paper, we make the following contributions. First, we propose multisite Chiron, with a novel architecture to execute SWfs in multisite cloud environments with provenance support. Second, we propose a novel multisite task scheduling algorithm, *i.e.* Data-Intensive Multisite task scheduling (DIM), for SWf execution in multisite Chiron. Third, we carry out an experimental evaluation, based on the implementation of multisite Chiron in Microsoft Azure using two SWfs, *i.e.* Buzz [8] and Montage [2].

This paper is organized as follows. Section 2 explains the design of a multisite SWfMS. Section 3 proposes our scheduling algorithm. Section 4 gives our experimental evaluation. Section 5 concludes the paper.

2   System Design

Chiron [15] is a data-centric SWfMS for the execution of SWfs at a single site, with provenance support. We extend Chiron to multisite, *i.e.* multisite Chiron, in order to manage the communication of Chiron instances at each site and automatically take advantage of distributed resources at each site to process the distributed data. In the execution environment of multisite Chiron, there is a master site (site 1 in Fig. 1) and several slave sites (Sites 2 and 3 in Fig. 1). The master site is composed of several Virtual Machines (VMs), a shared file system and a provenance database. The synchronization among different sites is achieved in a master-worker fashion while the intersite data transferring is realized in peer-to-peer. A slave site is composed of a cluster of VMs with a deployed shared file system. In the multisite environment, each VM is a computing node (or node for short). A node is selected as a master node at each site. In this paper, we assume that there is a Web domain at each site for SWf execution and that all the resources related to the SWf execution are in the Web domain.

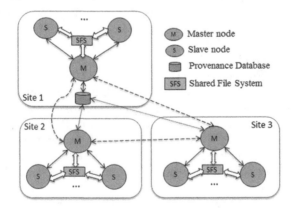

Fig. 1. Architecture of multisite Chiron.

First, Chiron analyzes the data dependencies of each activity. When the input data of an activity is ready [15], it generates tasks. Then, the tasks of each activity are independently scheduled to each site. All the previous processes are realized at the master node of the master site. Then, the data is transferred to the scheduled sites and the tasks are executed at the scheduled sites. Although the input data of a SWf cannot be moved, the intermediate data can be moved. The intermediate data is the data generated by processing the input data, which is the input data of following activities. After the execution of tasks, the provenance data [15] of each site are transferred to the provenance database. When the tasks of all the activities are executed, the execution of a SWf is finished.

Provenance data contains the information on activities, tasks and sites. An activity has an operator, *i.e.* the program for this activity and its status can be ready, running or finished. The start and end times of an activity are also recorded. An activity is related to several tasks, input relations and output relations. Each relation has its own attributes and tuples. The tasks of an activity are generated based on the input relation. A task processes the files associated with the corresponding activity. Each task has also a status, *i.e.* ready, running or finished and its start and end times are also recorded. The information about sites, *e.g.* computing capacity, files stored and tasks executed at the sites, are also stored as provenance data. The provenance data can be stored at each site first and then transferred to the centralized provenance database asynchronously with the execution of other tasks. In long running SWfs, provenance data needs to be queried at runtime for monitoring. In a multisite environment, the provenance data transfer may have a major influence on the data transfer of task scheduling. Latency hiding techniques can be used to hide the time to transfer data but it is difficult to hide the time to transfer the real-time provenance data generated during execution. Overall, the multisite scheduling problem should take into account the resources at different sites and intersite data transfer, including intermediate data to be processed by tasks and the provenance data.

3   Task Scheduling

In this section, we propose our two Level (2L) scheduling approach (see Fig. 2) for multisite execution of SWfs and a multisite scheduling algorithm, *i.e.* DIM. In the 2L scheduling approach, the first level performs multisite scheduling, where each task is scheduled to a site. DIM works at this level. The second level performs single site scheduling, where each task is scheduled to a VM by the default scheduling strategy (dynamic FAF [15]) of Chiron. A task is the assignment of an input data element to its corresponding activity to be executed. When an activity has n input data elements n tasks are executed independently. Synchronization is based on data dependency between activities as defined at the SWf specification. In these experiments we forced a synchronization so that the next activity only starts after all tasks of the previous activity are executed. In this paper, we focus on the first level, *i.e.* we consider scheduling tasks to different sites while the process of scheduling tasks to each VM is realized by the default scheduling approaches within the original SWfMS. The method to estimate the total time (including the time to transfer intersite data) to execute a bag of tasks at a single site is detailed in Sect. 3.2.

The DIM algorithm is shown in Algorithm 1. First, the tasks are scheduled according to the location of input data (Lines 2–5), which is similar to the

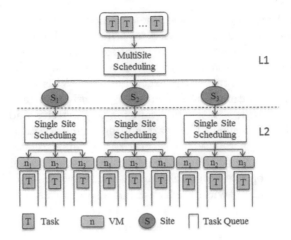

Fig. 2. MultiSite Scheduling. The master node at the master site schedules tasks to each site. At each site, the master node schedules tasks to slave nodes.

scheduling algorithm of MapReduce [4]. Line 3 searches Site s that stores the biggest part of input data corresponding to task t. Line 4 schedules Task t at Site s. The scheduling order (the same for Algorithm 2) is based on the id of each task. Line 5 estimates the total time to execute all the tasks scheduled at Site s according to Formula 3.2.2, considering the time for intersite data transfer. Then, the total time at each site is balanced by adjusting the whole bag of tasks scheduled at that site (Lines 6–9). Line 6 checks if the maximum difference of the estimated total time to execute tasks at each site can be reduced by verifying if the difference is reduced in the previous loop or if this is the first loop. While the maximum difference can be reduced, the tasks of the two sites are rescheduled as described in lines 7–9. Lines 7 and 8 choose the site that has the minimal total time and the site that has the maximum total time, respectively. Then, the scheduler calls the function $TaskReschedule$ to reschedule the tasks scheduled at the two selected sites to reduce the maximum difference of total time.

In order to achieve load balancing of two sites, we propose $TaskReschedule$ algorithm. Let us assume that there are two sites, $i.e.$ Sites s_i and s_j. For the tasks scheduled at each site, we assume that the total time at Site s_i is bigger than Site s_j. In order to balance the total time at Sites s_i and s_j, some of the tasks scheduled at site s_i should be rescheduled at Site s_j. In Algorithm 2, Line 1 calculates the difference of the total time at two sites according to Formula 3.2.2 with a scheduling plan. Line 2 gets all the tasks scheduled at site s_i. For each Task t in T_i (line 3), it is rescheduled at Site s_j if the difference of total time at the two sites can be reduced (lines 4–8). The task that has no input data at Site s_j is rescheduled first. Line 4 reschedules Task t at Site s_j. Line 5 calculates the total time at Sites s_i and s_j. Lines 6–7 update the scheduling plan if it can reduce the difference of total time at the two sites by rescheduling Task t.

Algorithm 1. Data-Intensive Multisite task scheduling (DIM)

Input: T: a bag of tasks to be scheduled; S: a set of cloud sites
Output: SP: the scheduling plan for T in S
1: $SP \leftarrow \emptyset$
2: **for each** $t \in T$ **do**
3:     $s \leftarrow GetDataSite(t)$
4:     $SP \leftarrow SP \cup \{Schedule(t,s)\}$
5:     EstimateTime(T, s, SP)
6: **while** MaxunbalanceTime(T, s, SP) is reduced in the last loop **do**
7:     $sMin \leftarrow MinTime(S)$
8:     $sMax \leftarrow MaxTime(S)$
9:     TaskReschedule($sMin$, $sMax$, SP)
end

Algorithm 2. Task Reschedule

Input: s_i: a site that has bigger total time for its scheduled tasks; s_j: a site that has smaller total time for its scheduled tasks; SP: original scheduling plan for a bag of tasks T
Output: SP: modified scheduling plan
1: $Diff \leftarrow CalculateExecTimeDiff(s_i, s_j, SP)$ ▷ Absolute value
2: $T_i \leftarrow GetScheduledTasks(s_i, SP)$
3: **for each** $t \in T_i$ **do**
4:     $SP' \leftarrow ModifySchedule(SP, \{Schedule(t, s_j)\}$
5:     $Diff' \leftarrow CalculateExecTimeDiff(s_i, s_j, SP')$
6:     **if** $Diff' < Diff$ **then**
7:         $SP \leftarrow SP'$
8:         $Diff \leftarrow Diff'$
end

3.1 Complexity

In this section, we analyze the complexity of the DIM algorithm without considering the local scheduling within each single cloud site. Let us assume that we have n tasks to be scheduled at m sites. The complexity of the first loop (lines 2–5) of the DIM algorithm is $\mathcal{O}(n)$. The complexity of the $TaskReschedule$ algorithm is $\mathcal{O}(n)$, since there may be n tasks scheduled at a site in the first loop (lines 2–5) of the DIM algorithm. Assume that the difference between maximum total time and minimum total time is T_{diff}. The maximum value of T_{diff} can be $n * avg(T)$ when all the tasks are scheduled at one site while there is no task scheduled at other sites. $avg(T)$ represents the average execution time of each task, which is a constant value. After m times of rescheduling tasks between the site of the maximum total time and the site of the minimum total time, the maximum difference of total time of any two sites should be reduced to less than $\frac{T_{diff}}{2}$. Thus, the complexity of the second loop (lines 6–9) of the DIM algorithm is $\mathcal{O}(m \cdot n \cdot \log n)$. Therefore, the complexity of the DIM algorithm is

$\mathcal{O}(m \cdot n \cdot \log n)$. It is only a little bit higher than that of OLB and MCT, which is $\mathcal{O}(m \cdot n)$, but yields high reduction in SWf execution (see Sect. 4).

3.2 Execution Time Estimation

We now give the method to estimate the total time to execute a bag of tasks at a single site, which is used in both the DIM algorithm and the MCT algorithm. Formula 3.2.1 gives the estimation of total time without considering the time to generate provenance data, which is used in the MCT algorithm.

$$
\begin{aligned}
TotalTime(T, s) = &ExecTime(T, s) \\
&+ InputTransTime(T, s)
\end{aligned}
\tag{3.2.1}
$$

T represents the bag of tasks scheduled at site s. $ExecTime$ is the time to execute the bag of tasks T at site s, $i.e.$ the time to run the corresponding programs. $InputTransTime$ is the time to transfer the input data of the tasks from other sites to site s. In the DIM algorithm, we use Formula 3.2.2 to estimate the total time with the consideration of the time to generate provenance data.

$$
\begin{aligned}
TotalTime(T, s) = &ExecTime(T, s) \\
&+ InputTransTime(T, s) \\
&+ ProvTransTime(T, s)
\end{aligned}
\tag{3.2.2}
$$

$ProvTransTime$ is the time to generate provenance data.

We assume that the workload of each task of the same activity is similar. The average workload (in GFLOP) of the tasks of each activity and the computing capacity of each VM at Site s is known to the system (configured by SWfMS users). The computing capacity (in GFLOPS) indicates the workload that can be realized per second, which can also be configured by users. Then, the time to execute the tasks can be estimated by dividing the total workload by the total computing capacity of Site s, as shown in Formula 3.2.3.

$$
ExecTime(T, s) = \frac{|T| * AvgWorkload(T)}{\sum_{VM_i \in s} ComputingCapacity(VM_i)}
\tag{3.2.3}
$$

$|T|$ represents the number of tasks in Bag T. $AvgWorkload$ is the average workload of the bag of tasks.

The time to transfer input data can be estimated as the sum of the time to transfer the input data from other sites to Site s as shown in Formula 3.2.4.

$$
InTransTime(T, s) = \sum_{t_i \in T} \sum_{s_i \in S} \frac{InDataSize(t_i, s_i)}{DataRate(s_i, s)}
\tag{3.2.4}
$$

$InDataSize(t_i, s_i)$ represents the size of input data of Task t_i, which is stored at Site s_i. The size can be measured at runtime. $DataRate(s_i, s)$ represents the data transfer rate, which can be configured by users. S represents the set of sites.

Finally, the time to generate provenance data is estimated by Formula 3.2.5.

$$ProvTransTime(T, s) = |T| * TransctionTimes(T)$$
$$* AvgTransactionTime(s) \tag{3.2.5}$$

$|T|$ represents the number of tasks in Bag T. We can estimate $AvgTransactionTime$ by counting the time to perform a data transfer to update the provenance data of a task in the provenance database from Site s. $TransctionTimes(T)$ represents the number of data transfers to perform for generating the provenance data of each task in Bag T. It can be configured according to the features of the SWfMS.

4   Experimental Evaluation

This section gives our experimental evaluation of DIM, using multisite Chiron within Microsoft Azure. Azure [1] has multiple cloud sites, *e.g.* Central US (CUS), West Europe (WEU) and North Europe (NEU). We instantiated three A4 (8 CPU cores, see details in [3]) VMs at each of the three site, *i.e.* CUS, WEU and NEU. We take WEU as a master site. We deploy an A2 VM (2 CPU cores) at WEU and install PostgreSQL database to manage provenance data. We evaluate DIM by executing Buzz [8] and Montage [2]. We assume that the input data of the SWfs are distributed at the three sites. We also assume that the scheduling of tasks to each VM (or CPU core) is done by the default scheduling approach of Chiron [15]. We compare our proposed algorithm with two representative baseline scheduling algorithms, *i.e.* Opportunistic Load Balancing (OLB) and Minimum Completion Time (MCT). In the experiment, we assume the input data of SWfs cannot be moved and schedule the tasks of the activity that processes the input data to where the data is while exploiting DIM, OLB or MCT to schedule tasks of the other activities. In the multisite environment, OLB randomly selects a site for a task while MCT schedules a task to the site that can finish the execution first. HEFT is a heuristic scheduling approach, which is commonly used for SWf scheduling. The difference between HEFT and MCT is that HEFT ranks the tasks according to the dependencies and workload of tasks before scheduling the tasks. However, since the tasks of one activity are independent of each other, HEFT degrades to MCT in our case. In Figs. 3 and 4, the execution time is the absolute time for SWf execution and the data-transfer size refers to the input data of tasks, *i.e.* the intermediate data transferred across different sites, which does not include the provenance data. In addition, since the resource utilization also depends on the programs used in different SWfs and that each SWf exploits various programs, we did not measure it.

First, we used a DBLP 2013 XML file of 60 MB as input data for Buzz SWf in our experiments. The input data is partitioned into three parts, which have almost the same amount of data, and each part is stored at a site while configuration files of Buzz SWf are present at all the three sites. The centralized provenance database is located at the master site. As shown in Fig. 3(a), DIM is

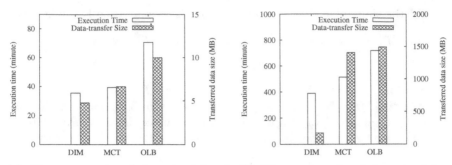

(a) The amount of input data is 60MB.

(b) The amount of input data is 1.29GB.

Fig. 3. Buzz SWf execution.

much better than MCT and OLB in terms of both execution time and transferred data size. The execution time corresponding to DIM is 9.6% smaller than that corresponding to MCT and 49.6% smaller than that corresponding to OLB. The size of the data transferred among different sites corresponding to MCT is 38.7% bigger than that corresponding to DIM and the size corresponding to OLB is 108.6% bigger than that corresponding to DIM.

Second, we performed an experiment using a DBLP 2013 XML file of 1.29 GB as input data for Buzz while configuration files of Buzz are present at all the three sites. The other configuration is the same as the first one. As shown in Fig. 3(b), the advantage of DIM in terms of both execution time and transferred data size compared with MCT and OLB increases with bigger amounts of input data. The execution time corresponding to DIM is 24.3% smaller than that corresponding to MCT and 45.9% smaller than that corresponding to OLB. The size of the data transferred between different sites corresponding to MCT is 7.19 times bigger than that corresponding to DIM and the size corresponding to OLB is 7.67 times bigger than that corresponding to DIM. Although OLB is a random algorithm, it distributes the tasks to each site with the same probability and the transferred data remains the same for the same configuration of the SWf and cloud environment. As a result, the size of intersite transferred data can represent the average results, which are calculated from the execution of multiple tasks.

Since DIM considers the time to transfer intersite provenance data and makes optimization for a bag of tasks, *i.e.* global optimization, it can reduce the total time. Since DIM schedules the tasks to where the input data is located at the beginning, DIM can reduce the amount of intersite transferred data compared with other algorithms. MCT only optimizes the load balancing for each task, *i.e.* local optimization, among different sites without consideration of the time to transfer intersite provenance data. It is a greedy algorithm that can reduce the execution time of a SWf by balancing the total time of each site while scheduling tasks of each activity. However, it cannot optimize the scheduling for the whole execution of all the tasks at each site. In addition, compared with OLB, MCT cannot reduce much the transferred data among different sites. Since OLB simply tries to keep all the sites working on arbitrary tasks, it has the worst performance.

Furthermore, we executed the Montage SWf of 0.5° with three sites, *i.e.* CUS, WEU and NEU. The size of input data is 5.5 GB. The input data is evenly partitioned to three parts stored at the corresponding sites with configuration files stored at all the three sites. The execution time and amount of intersite transferred data are shown in Fig. 4(a).

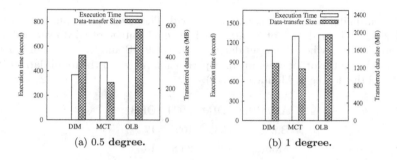

Fig. 4. Montage SWf execution.

The execution results of Montage with 0.5° [6] reveals that the execution time corresponding to DIM is 21.7% smaller than that of MCT and 37.1% smaller than that corresponding to OLB. This is expected since DIM makes optimization for a bag of tasks in order to reduce intersite transferred data with consideration of the time to transfer intersite intermediate data and provenance data. MCT is optimized for load balancing only with the consideration of input data of tasks. OLB has no optimization for load balancing. In addition, the intersite transferred data of DIM is 42.3% bigger than that of MCT. Since DIM is designed to achieve load balancing of each site to reduce execution time, it may yield more intersite transferred data in order to achieve load balance. However, the amount of intersite transferred data of DIM is 28.6% smaller than that of OLB. This shows the efficiency of the optimization for the data transfer of DIM. Moreover, when the degree (0.5) is low, there is less data to be processed by Montage, and the number of tasks to schedule is small. Since DIM is designed for big numbers of tasks, the amounts of intersite transferred data are not reduced very much in this situation.

Finally, we executed Montage SWf of 1° in the multisite cloud. We used the same input data as in the previous experiment, *i.e.* 5.5 GB input data evenly distributed at three sites. The execution time and the amount of intersite transferred data are shown in Fig. 4(b).

The execution results of Montage with 1° reveals that the execution time of DIM is 16.4% smaller than that of MCT and 17.8% smaller than that of OLB.

As explained before, this is expected since DIM can reduce the execution time by balancing the load among different sites compared with MCT and OLB. In addition, the intersite transferred data of DIM is 10.7% bigger than that of MCT. This is much smaller than the value for 0.5° (42.3%), since there are more tasks to schedule when the degree is 1 and DIM reduces intersite transferred data for a big amount of tasks. However, the amount of intersite transferred data is bigger than that of MCT. This happens since the main objective of DIM is to reduce execution time instead of reducing intersite transferred data. In addition, the amount of intersite transferred data of DIM is 33.4% smaller than that of OLB, which shows the efficiency of the optimization for the data transfer of DIM.

Table 1. Scheduling time. The time unit is second. The size of the input data of Buzz SWf is 1.2 GB and the degree of Montage is 1 (The advantage of DIM over MCT and OLB is more obvious when the input data of Buzz is 1.2 GB and the degree of Montage is 1 compared with the other cases in our experiments, *i.e.* when the input data of Buzz is 60 MB and the degree of Montage is 0.5.).

Algorithm	DIM	MCT	OLB
Buzz	633	109	17
Montage	29.2	28.8	1.5

In addition, we measured the time to execute the scheduling algorithms to generate scheduling plans. The time is shown in Table 1. The complexity of MCT is the same as that of OLB, which is $\mathcal{O}(m \cdot n)$. However, the scheduling time of MCT is much bigger than OLB. The reason is that MCT needs to interact with the provenance database to get the information of the files in order to estimate the time to transfer the files among different sites. The table shows that the time to execute DIM is much higher than OLB for both Buzz and Montage since the complexity of DIM is higher than that of OLB and that DIM has more interactions with the provenance database in order to estimate the total time to execute the tasks at a site. When there are significant number of tasks to schedule (for the Buzz SWf), the time to execute DIM is much bigger than that of MCT because of higher complexity. However, when the number of tasks is not very big, the time to execute DIM is similar to that of MCT. The time to execute DIM and MCT is much bigger than that of OLB, since it takes much time to communicate with the provenance database for the estimation of the total time of each site. The scheduling time of the three scheduling algorithms is always small compared with the total execution (less than 3%), which is acceptable for the task scheduling during SWf execution. Although the scheduling time of DIM is much bigger than MCT and OLB, the execution time of SWfs corresponds to DIM is much smaller than that of MCT and OLB as explained in the four experiments. This means that DIM generates better scheduling plans compared with MCT and OLB.

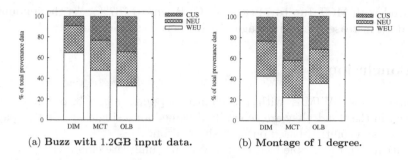

(a) **Buzz with 1.2GB input data.**   (b) **Montage of 1 degree.**

Fig. 5. The distribution of provenance during the execution of SWfs.

Furthermore, we measured the size of provenance data and the distribution of the provenance data. As shown in Table 2, the amount of the provenance data corresponding to the three scheduling algorithms are similar (the difference is less than 8%). However, the distribution of the provenance data is different. In fact, the bandwidth between the provenance database and the site is in the following order: WEU > NEU > CUS [1]. As shown in Figs. 5(a) and (b), the provenance data generated at CUS site is much more than that generated at NEU site and WEU site for DIM algorithm. In addition, the percentage of provenance data at WEU corresponding to DIM is much bigger than MCT (up to 95% bigger) and OLB (up to 97% bigger). This indicates that DIM can schedule tasks to the site (WEU) that has bigger bandwidth with the provenance database (the database is at WEU site), which yields bigger percentage of provenance data generated at the site. This can reduce the time to generate provenance data in order to reduce the overall multisite execution time of SWfs. However, MCT and OLB is not sensitive to the centralized provenance data, which correspond to bigger execution time.

Table 2. Size of provenance data. The data unit is MB. The size of the input data of Buzz SWf is 1.2 GB and the degree of Montage is 1.

Algorithm	DIM	MCT	OLB
Buzz	301	280	279
Montage	10	10	10

From the experiments, we can see that DIM performs better than MCT (up to 24.3%) and OLB (up to 49.6%) in terms of execution time although it takes more time to generate scheduling plans. DIM can reduce the intersite transferred data

---

[1] For instance, the time to execute "SELECT count(*) from eactivity" at the provenance database from each site: 0.0027s from WEU site, 0.0253s from NEU site and 0.1117s from CUS site.

compared with MCT (up to 719%) and OLB (up to 767%). As the amount of input data increases, the advantage of DIM becomes more important.

5 Conclusion

Although some SWfMSs with provenance support, *e.g.* Chiron, have been deployed in the cloud, they are generally designed for a single site. In this paper, we proposed a solution based on multisite Chiron.

Multisite Chiron is able to execute SWfs in a multisite cloud with geographically distributed input data. We proposed the architecture of multisite Chiron with a centralized provenance database. Based on this architecture, we proposed a new scheduling algorithm, *i.e.* DIM, which considers the latency to transfer input data of tasks and to generate provenance data in a multisite cloud. We analyzed the complexity of DIM ($\mathcal{O}(m \cdot n \cdot \log n)$), which is quite acceptable for scheduling bags of tasks. We used two real-life SWfs, *i.e.* Buzz and Montage to evaluate the DIM algorithm in Microsoft Azure with three sites. The experiments show that although its complexity is higher than that of OLB and MCT, DIM is much better than two representative baseline algorithms, *i.e.* MCT (up to 24.3%) and OLB (up to 49.6%), in terms of execution time. In addition, DIM can also reduce significantly (up to 7 times) the data transferred among different sites, compared with MCT and OLB. The advantage of DIM becomes important with big numbers of tasks.

Acknowledgment. Work partially funded by EU H2020 Programme and MCTI/RNP-Brazil (HPC4E grant agreement number 689772), CNPq, FAPERJ, and INRIA (MUSIC project), Microsoft (ZcloudFlow project) and performed in the context of the Computational Biology Institute (www.ibc-montpellier.fr). We would like to thank Weiwei Chen and Pegasus project for the help in modeling and executing the Montage SWf.

References

1. Microsoft Azure. http://azure.microsoft.com
2. Montage. http://montage.ipac.caltech.edu/docs/gridtools.html
3. Parameters of different types of vms in microsoft Azure. https://azure.microsoft.com/en-us/pricing/details/virtual-machines/
4. Dean, J., Ghemawat, S.: Mapreduce: simplified data processing on large clusters. In: 6th Symposium on Operating System Design and Implementation (OSDI), pp. 137–150 (2004)
5. Deelman, E., Gannon, D., Shields, M., Taylor, I.: Workflows and e-science: an overview of workflow system features and capabilities. Future Gener. Comput. Syst. **25**(5), 528–540 (2009)
6. Deelman, E., Singh, G., Livny, M., Berriman, B., Good, J.: The cost of doing science on the cloud: the montage example. In: International Conference for High Performance Computing, Networking, Storage and Analysis, pp. 1–12 (2008)

7. Deelman, E., Singh, G., Su, M.-H., Blythe, J., Gil, Y., Kesselman, C., Mehta, G., Vahi, K., Berriman, G.B., Good, J., Laity, A., Jacob, J.C., Katz, D.S.: Pegasus: a framework for mapping complex scientific workflows onto distributed systems. Sci. Program. **13**(3), 219–237 (2005)

8. Dias, J., Ogasawara, E.S., de Oliveira, D., Porto, F., Valduriez, P., Mattoso, M.: Algebraic dataflows for big data analysis. In: IEEE International Conference on Big Data, pp. 150–155 (2013)

9. Duan, R., Prodan, R., Li, X.: Multi-objective game theoretic scheduling of bag-of-tasks workflows on hybrid clouds. IEEE Trans. Cloud Comput. **2**(1), 29–42 (2014)

10. Etminani, K., Naghibzadeh, M.: A min-min max-min selective algorihtm for grid task scheduling. In: The Third IEEE/IFIP International Conference in Central Asia on Internet (ICI 2007), pp. 1–7 (2007)

11. Liu, J., Pacitti, E., Valduriez, P., de Oliveira, D., Mattoso, M.: Multi-objective scheduling of scientific workflows in multisite clouds. Future Gener. Comput. Syst. **63**, 76–95 (2016)

12. Liu, J., Pacitti, E., Valduriez, P., Mattoso, M.: A survey of data-intensive scientific workflow management. J. Grid Comput. **13**(4), 1–37 (2015)

13. Liu, J., Silva, V., Pacitti, E., Valduriez, P., Mattoso, M.: Scientific workflow partitioning in multisite cloud. In: Lopes, L., et al. (eds.) Euro-Par 2014. LNCS, vol. 8805, pp. 105–116. Springer, Cham (2014). doi:10.1007/978-3-319-14325-5_10

14. Maheswaran, M., Ali, S., Siegel, H.J., Hensgen, D., Freund, R.F.: Dynamic matching and scheduling of a class of independent tasks onto heterogeneous computing systems. In: 8th Heterogeneous Computing Workshop, p. 30 (1999)

15. Ogasawara, E.S., Dias, J., Silva, V., Chirigati, F.S., de Oliveira, D., Porto, F., Valduriez, P., Mattoso, M.: Chiron: a parallel engine for algebraic scientific workflows. Concurr. Comput. Pract. Exp. **25**(16), 2327–2341 (2013)

16. Pineda-Morales, L., Costan, A., Antoniu, G.: Towards multi-site metadata management for geographically distributed cloud workflows. In: 2015 IEEE International Conference on Cluster Computing, (CLUSTER), pp. 294–303 (2015)

17. Smanchat, S., Indrawan, M., Ling, S., Enticott, C., Abramson, D.: Scheduling multiple parameter sweep workflow instances on the grid. In: 5th IEEE International Conference on E-Science, pp. 300–306 (2009)

18. Topcuouglu, H., Hariri, S., Wu, M.: Performance-effective and low-complexity task scheduling for heterogeneous computing. IEEE Trans. Parallel Distrib. Syst. **13**(3), 260–274 (2002)

19. Wieczorek, M., Prodan, R., Fahringer, T.: Scheduling of scientific workflows in the ASKALON grid environment. SIGMOD Rec. **34**(3), 56–62 (2005)

Aspect Oriented Parallel Framework for Java

Bruno Medeiros and João L. Sobral[✉]

Departamento de Informática, Universidade do Minho, Braga, Portugal
{brunom,jls}@di.uminho.pt

Abstract. This paper introduces aspect libraries, a unit of modularity in parallel programs with compositional properties. Aspects address the complexity of parallel programs by enabling the composition of (multiple) parallelism modules with a given (sequential) base program. This paper illustrates the introduction of parallelism using reusable parallel libraries, coded in AspectJ. These libraries provide performance comparable to traditional parallel programming techniques and enable the composition of multiple parallelism modules (e.g., shared memory with distributed memory) with a given base program.

1 Introduction

OpenMP and MPI are arguably the most relevant instances of the shared memory (SM) and distributed memory (DM) parallel programming paradigms (PPP), respectively. With the increase of clusters of multicore machines it is common to combine MPI with OpenMP to provide a hybrid solution. However, parallelism related concerns (PRC) are known for being crosscutting concerns (CCC) [1], so it is frequent to mix them up with domain application concerns, jeopardizing the application maintenance and evolution. The situation is even more grotesque in some low level parallel programming codes where performance is the primary and virtually the exclusive goal. This mixing up of concerns is known by tangling and scattering [16]. The former refers to a chunk of code (e.g., class) that implements more than one concern (e.g., PRC and domain concerns), whereas the latter refers to a concern that is spread out over multiple modules. Although OO promotes and offers means to modularise concerns, and arguably has better mechanisms to do so than low level languages such as C, it falls short to modularise CCC (e.g., parallelism). Furthermore, most parallel programming languages (e.g., OpenMP) provide high-level abstractions to deal with the parallelism requirements of a specific programming model. Hence, to take advantage of hierarchical environments, such as clusters of multicores, it might be necessary to program with two different parallel programming languages and combine them together. This not only increases the complexity of software design, learning curve, and likelihood of bugs but also leads to even more code tangling and scattering issues.

Figure 1 presents the most time consuming function of a molecular dynamic simulation, the force calculation among particles. This function is composed by two nested cycles, one outer cycle for all the system particles (line 04) and one

I. Dutra et al. (Eds.): VECPAR 2016, LNCS 10150, pp. 220–233, 2017.
DOI: 10.1007/978-3-319-61982-8_20

```
00:  void MD (..){
01:  ...
02:  forces = particles.getForces();
03:
04:    for (pA = 0; pA <  maxParticles; pA ++)
05:     for(pB = pA + 1; pB < maxParticles; pB++)
06:      if(distance(pA, pB) < radius){
07:         forcesAB = callForcesParticles(pA,pB);
08:         forces[pA] += forcesAB;
09:         forces[pB] -= forcesAB;      // Newton's 3rd Law
10:      }
11:} ...
```

Fig. 1. Molecular dynamic - force calculation among particles.

inner cycle, varying from the current particle to the total number of particles in the system (line 05). For each inner cycle iteration the force between pairs of particles, in a given radius (line 06), is updated, based on particles' positions. Notice that this algorithm uses the third Newton's law (line 09), which states that the force that a body A applies in B, in module, is equal to the force that B applies in A. So whenever two particles are in the same radius, both particles' forces are update, therefore reducing the number of iterations from $N(N-1)^2$ to $\frac{N(N-1)}{2}$. Although this optimisation reduces the sequential execution time to half, it also makes future parallel versions harder to implement, since it imposes both load balancing and mutual exclusion challenges that should be dealt with.

With a more in-depth analysis of the code from Fig. 1 it is possible to identify the outer loop as a hot spot to be parallelized and that in a SM parallelisation there would be a risk of race condition during the update of particles' forces.

Figure 2 presents a hybrid parallelisation of the code from Fig. 1 to illustrate the tangling and scattering issues. The iterations of the outer loop were divided by all threads of all processes (line 04) and since the forces' updates might cause data races, each thread has its own local array to save the particles' forces (lines 08 and 09). In the end threads call a local barrier and reduce their work to the master thread (line 12). Lastly, the master thread of each process performs a force reduction among them (line 15). The black, blue and red lines of code are related with the sequential, multi-thread and multi-process concerns, respectively.

Figure 2 shows that adding parallelism into the sequential code made it more complex and harder to understand. If the developer wants to change the mapping between particles and threads/processes to improve the load balancing, he has to rewrite the code. Moreover, if the optimisations are duplicated elsewhere this means that any modification to its reasoning provokes code changes in different locations of the application. This exposes the inherent problem of the lack of modularity of such solutions. A better design solution would be to encapsulate threads and processes parallelisation into independent modules.

```
00:  void MD (..){
01:  ...
02:  globalID = threadID + processID * numThreads;
03:  totalWorkers = numProcess * numThreads
04:    for (pA = globalID ; pA < sizeP; pA += totalWorkers)
05:     for(pB = pA + 1; pB < sizeP; pB++)
06:      if(distance(pA, pB) < radius){
07:         forcesAB = callForcesParticles(pA,pB);
08:         threadForces[threadID][pA] += forcesAB
09:         threadForces[threadID][pB] -= forcesAB
10:      }
11:  callThreadBarrier();
12:  threadForceReduction();
13:  callThreadBarrier();
14:  if (threadID == masterThread)
15:       processForceReduction();
16:} ...
```

Fig. 2. MD - Hybrid parallelisation (Threads + Processes). (Color figure online)

Dealing with the thread parallelisation by using OpenMP annotations in the example of Fig. 2 would reduce some of the its unreadability. However, those annotations are still scattered and tangled with the sequential code. Furthermore, annotations can only be used to provide the basic parallel constructs (e.g., parallel region). For example, the statements related with thread local arrays (lines 08 and 09) would still remain, as well as processes parallelisation concerns. In a Java context, a more OO approach could have been used, such as moving the PRC into a base class, which would extend the target class. However, in this approach the target class can not extend another class.

This paper shows an approach to develop high performance hybrid parallel Java applications without polluting its source code. Initially Java programmers develop their sequential base code and further on add the parallelism modules to it. These modules are added in a non-invasive fashion and their PRC are inserted at compile/load time. The modules are pluggable, which allows testing different types of parallelism without rewriting the base code every time. Performance portability is addressed by supporting SM and DM libraries that can work together or separated to address the specificities of each target platform.

The next section discusses how our parallel libraries address the PRC and presents their implementation that mimics the current mainstream PPP. Section 3 presents performance results. Section 4 compares this work with related work. Finally, Sect. 5 concludes the paper.

2 Parallel Libraries with AspectJ for Hybrid Parallelism

To solve the problem of PRC we provide SM and DM libraries in AspectJ to be used as an extra layer of modularisation, promoting a modular approach.

AspectJ [3] is a solution to provide modules that can be added to the base program without polluting it with CCC. With AspectJ it is possible to modify the static structure of an application (e.g., class hierarchy), as well as its execution flow in a non-invasive fashion. This language allows the capture of join points (using pointcuts), spread across a base program, to add behavior to them (using advices). This behavior is explicitly added (e.g., at compile/load time) through code transformations performed by AspectJ's internal mechanisms, providing a solution to deal with the tangled and scattering problems. Since multiple transformations can be applied to the same base program, the language allows to specify hybrid parallelisations. However, AspectJ restricts the granularity and the type of join points that can be triggered. Therefore, some base programs might need to be adapted to expose potential join points.

Our approach identifies three main components, the base program, the aspect libraries and the concrete aspect, represented in the Fig. 3 by blue, red and black colors, respectively. Each library is represented by an abstract aspect that is connected with external APIs (e.g. Java Threads, OpenMPI ...) in a reusable and decoupled manner. Thus, programmers can easily interchange between different external APIs. Furthermore, since the libraries will be used with different applications, they cannot depend on join points of a specific application. Thus, these libraries are composed by abstract aspects that encapsulate behavior and state transversal to their sub-aspects and abstract pointcuts without explicitly defined join points. Later on, for each application, those abstract aspects are extended by concrete ones that encapsulate state and behavior specific to the target application. The mapping between the abstract pointcut and the join points to be intercepted is defined in the concrete aspect. The concrete aspect works as a bridge between the core of the library of aspects and the target application, resembling the use of XML configuration files in the Spring AOP framework [13]. Finally, from the application point of view, in some cases it is necessary to expose join points using our design rules. This kind of approach is known in requirements engineering as *scaffolding* [14].

The SM library is influenced by OpenMP and provides many of its constructs, such as: critical, single, master, barrier, parallel for (dynamic, static ...), tasks and so on. It is possible to specify how objects behave among threads, allowing to declare them as private, to be reduced and so on. The SM library uses a threads-executors pool created after intercepting the main method of the target application. Whenever a thread reaches a parallel region requests from the initial pool a new team of threads and becomes the master this team.

The DM library will run as many instances of the base program as the number of processes requested using the SPMD execution model of MPI. This library provides constructs implemented on top of MPI calls, thus offering constructs such as: - Allreduce, gather, scatter, broadcast and many others. Moreover, offers constructs that are not provided by the MPI standard, for example: parallel for (with static round-robin, static by blocks and dynamic distributions) and distribution of 2D arrays using several strategies.

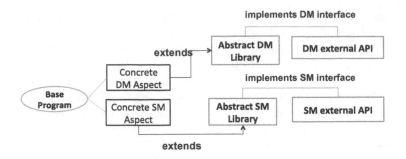

Fig. 3. Aspect libraries: overview. (Color figure online)

Along with the parallel libraries, it is also provided a data API with extra features, such as: different types of scatter of a matrix among processes (e.g., by lines and so on); accessing arrays with high-level abstractions (e.g., gets/sets ...); pre-programmed reduction functions for arrays and matrices and so on.

Similar to OpenMP and MPI, our libraries do not check for data dependencies, race conditions or deadlocks. Nevertheless, the libraries guarantee the correctness of its aspects and advices, and of the user concrete aspects as long as it follows our designing rules. As far as AspectJ is concerned, with our approach the user only specifies well defined pointcuts and/or inter-type declaration. Thus, reducing the complexity of using our approach and facilitating its correctness.

2.1   Design Rules

AspectJ uses the concept of join point which refers to any identifiable point in the execution control flow. However, for designing reasons, the language restricts the granularity and the types of the join points that can be trigger. Limiting the access to more stable constructs allows to better control the complexity, as well as potential harmful side effects of using a broader join point model. The AspectJ join point model is a well defined one [13] that includes among others the call/execution of a method, object/class initialisation, set and get of object fields and so. However, this model does not include join points such as the interception of local variables, the body of a loop *for*, the accessed position of a given array and so forth. When a programmer wants to inject additional behavior into a point in the source code that is not part of the AspectJ join point model, all he needs to do is to transform such point into an identifiable AspectJ join point. Most of the times a creation of a method that encapsulates the desirable point is enough. We offer a set of design rules that will help the user to identify and deal with the situation where transformation to the source code should be done in order to highlight potential parallel related join points.

In our approach domain experts develop sequential code and apply, if necessary, soft design rules that enable the introduction of PRC and are a key to enable the composition of parallelism modules. Those design rules are the same to every application and work in compliance with our aspect libraries.

Our first design rule states that PRC should be encapsulated around methods. In this manner, PRC can be uniquely identified and additional behavior can be easily (un)plugged. Additionally in some cases it is necessary to expose some context of the desirable join point using the arguments of the newly created method. For example, if the desirable join point is a loop that we want to parallelize the iteration range of such loop should be passed as arguments of the created method. Performance-wise, since such methods can be declared as final, the compiler will most-likely inline its calls.

The above method design rule, when applied to loops, can be formalized as follows: Consider a given cycle *for* where a represents the number of the first iteration, b the number of the last iteration, c the incremental step and Bc the block of code that will be executed inside the loop. The programmer must create a new method M (the name is defined by the programmer) where the first three arguments are a, b and c, in that order, and the remaining are all the local variables used by the Bc. The new method will contain the original *for* executing the Bc. However, the original values of the *for* are replaced by the names given to the variables a, b and c passed as the M arguments. Furthermore, the original loop *for* is replaced by the call to M, passing the values corresponding to a, b, c and the values of the remaining necessary variables (if there is any).

AspectJ can only identify variables of instance, therefore our rule to data states that: All the variables to become private, reducible or sendable across processes have to be variables of instance of an object. Furthermore, the class holding such variables have to implement our interfaces. Those interfaces are used as marker interfaces to identify objects that our libraries should intercept and perform actions (e.g., reductions). The declaration that an object implements our interfaces is coded in the concrete aspect instead of the target object. In this manner objects are not polluted with PRC. This is possible using the inter-type declaration mechanism of AspectJ. In SM this design rule is applied to objects that require to become private to threads, whereas in DM is applied to objects used in data communication among processes.

2.2   Illustrative Example

In Fig. 4 the lines 02, 05 and 08 illustrate the use of the loop design rule in the sequential code of Fig. 1. The user already dealt with the parallel task, now he needs to deal with the mutual exclusion. Lets say the user does not know yet if he should (1) synchronize the entire force update (line 12 and 13), (2) have a lock *per* particle or (3) use local arrays and perform a reduction in the end (following the same strategy of Fig. 2). The first approach uses the critical constructor of the library and implies the transformation of the block of code of line 12 and 13 into a method. Figure 5 illustrates such transformation.

The second approach (lock *per* particle) would not require any modification to the code of Fig. 5, since a method (*forceUpdate*) was already created and the access to the array positions is exposed in the arguments (*pA* and *pB*). Finally, the third approach also does not require modifications to the source code, since the forces are variables of instance of the object particles. Nevertheless, according

```
01: void MD (..) {
02:     forceCalculation(0, maxParticles, 1, ...);
03:}
04: ...
05: void forceCalculation(int begin, int end, int step, ...) {
06: ...
07:   forces = particles.getForces();
08:   for (pA = begin; pA < end; pA +=step)
09:     for(pB = pA + 1; pB < maxParticles; pB++)
10:      if(distance(pA, pB) < radius){
11:         forcesAB = callForcesParticles(pA,pB);
12:         forces[pA] += forcesAB;
13:         forces[pB] -= forcesAB;      // Newton's 3rd Law
14:      }
15:} ...
```

Fig. 4. MD - parallel for transformation

```
01: void MD (..) {
....
05: void forceCalculation (....) {
06: ...
07:   forces = particles.getForces();
08:   ...
12:   forceUpdate(pA, pB, forcesAB);
13: ...
14:} ...
15:   private void forceUpdate(... pA, ... pB, ... forcesAB){
16:       forces[pA] += forcesAB;
17:       forces[pB] -= forcesAB;      // Newton's 3rd Law
18: }
```

Fig. 5. MD - transformation for critical section

to the rules the user needs to specify that particles implement our interface. This action will be coded directly in the aspect instead.

The design rules that were applied above are also valid for the distributed memory library, namely the parallel for method (line 07 of Fig. 4) that will be used to distribute the iterations of the outer loop among processes as well as the join point to perform the forces reduction among processes.

Figure 6 presents the concrete aspects with the join points that will be intercepted to add the requested behavior by DM and SM libraries. In the concrete aspects the programmer expressed the intentions: - of statically dividing the outer loop iterations within method *forceCalculation()* among processes (line 11 of Fig. 6) and among their threads (line 19) and at the end of it performing a

```
00: public aspect parallel{
01:
02: pointcut parallelRegion():(call(.. forceCalculation(...)))
03: pointcut data()            :(call(... getForces()));
04: pointcut forParallel(...)    :
05:  (execution (void forceCalculation(int, int, int..)))...
06:
07: static aspect DM_Concrete extends   abstract_DM_Library {
08:
09:    declare particles.forces implements DMInterface
10:    ...
11:    pointcut staticFor(...) : forParallel(...)
12:    pointcut reduction()    : parallelRegion ();
13:    pointcut commData()     : data();
14: }
15: ...
16: static aspect SM_Concrete extends abstract_SM_Library {
17:    declare particles.forces implements SMInterface
18:    ...
19:    pointcut staticFor (...)    : forParallel(...);
20:    pointcut reduction()        : parallelRegion();
21:    pointcut privateData()      : data();
22: } ...
23:}
```

Fig. 6. Distributed memory and shared memory concrete aspects.

data reduction among threads (line 20) and among processes (line 12); - that particles' forces are objects that will became private (line 17) and used during processes communication (line 09) - that the *getForces* method will return a private thread copy (line 21) and that this data will be used in processes communication (line 13) as well. As we can see by the shared memory aspect the user opted to deal with mutual exclusion by using private data instead of the critical or lock *per* particle approaches.

In the Fig. 6 the SM and DM libraries work together to provide a hybrid parallel solution to the sequential code of Fig. 5. Nevertheless, if the SM aspect was commented the DM library would work alone providing a distributed memory parallelisation and *vice versa*. In the hybrid example of Fig. 6, after intercepting the main method, the DM and SM libraries will create data related to the processes and their pool of threads, respectively. Since, the object particles implements the SM and DM interfaces, the SM library will create a copy of the particles' forces for each thread and the DM library will save a reference of particles' forces of the master thread. Before entering the *forceCalculation()* method the DM library will intercept its arguments and modify them in order to assign the iterations of its loop (line 08 of Fig. 4) to the processes. The SM library will then further divide those iterations by the threads. When the *getForces()* method is intercepted the SM library will caught its object reference, match this reference in a internal *hashmap* and return the correspondent copy assigned to the current thread. After the *forceCalculation()* method finishes its work, the SM library will internally reduce all the forces among threads and save its result in the reference to the particles' forces object corresponding to the thread master. Finally, the DM library will perform, among processes, a global reduction of each master thread result.

To better understand how the parallel for's are composed both by the DM and in SM libraries lets say the user chooses a static distribution with chunk $= 1$ for both libraries and executes the simulation with 2 processes each with 2 threads. For each process the DM library will intercept the $forceCalculation(0, maxParticles, 1, ...)$ and change the arguments accordingly to the process id. The process 0 will run $forceCalculation(0, maxParticles, 2, ...)$ and the process 1 will run $forceCalculation(1, maxParticles, 2, ...)$, in another words the process 0 will go through the even iterations while the process 1 over the odd iterations. Then each thread created by the SM library will intercept the $forceCalculation$ of the their parent process. Thus thread 0 and thread 1 of process 0 will change the arguments of $forceCalculation$ to $forceCalculation(0, maxParticles, 4, ...)$ and $forceCalculation(2, maxParticles, 4, ...)$, respectively. Finally, the thread 0 and thread 1 of process 1 will execute $forceCalculation(1, maxParticles, 4, ...)$ and $forceCalculation(3, maxParticles, 4, ...)$, respectively.

2.3   Discussion

The strongest points of our approach are the separation of the CCC from the source code, centralizing such concerns into modularity units called aspects, sequential semantic and the requirement of learning only one new syntax (AspectJ). Typically, when parallelizing with Java threads, one has to either declare that a class extends the Thread class, overwriting the method *run*, or to make a given class to implement the Runnable *interface*, following some restrictions of the Java thread model (e.g. final variables). When the programmer wants to test multiple SM strategies (e.g., different loop distributions) normally he writes different versions connected to the same source code with a mechanism to choose between them (e.g., if and else). For example, in the code of Fig. 4 there are two PRC that should be dealt, namely how the iterations of the outer loop should be mapped to the threads and how to deal with the race condition. During the development process the programmer would probably develop different versions of the same code to test which best fits. Moreover, there might be different strategies that work better with different machines, so different strategies in the same code should be maintained, and so on. Things become even more complex if the programmer then wants to build a DM version and merge it with the SM version to build a hybrid version, making the code much more complex, unreadable and more error-prone. However, with our libraries, testing different approaches means just changing different pointcuts (as long as the design rules where followed). For example, in the Fig. 6 if the user wants to test a dynamic SM distribution he could change the pointcut from *staticFor* to *dynamicFor* and

the same holds true for the DM for distribution as well. This flexible testing approach increases productivity, reduces the introduction of bugs and improves the
code evolution.

The main disadvantages of our approach are the design rules and the code
correctness. However, most IDE (e.g., eclipse) not only help the user with code
refactoring (e.g., method refactoring) but also highlight the code points where
the pointcuts are being injected. The design rules are simply method refactoring,
which not only does not interfere with code evolution but also frequently makes
the code more readable. The unpleasant part of this refactoring is the variables
that should be passed as arguments, but this is also attenuated with IDE tools.
Concerning the data design rules, aside from some exceptions, most of the time
relevant variables are already variables of instance. Also, our design rule is similar
with the way programmers deal with data using Java threads. A secondary
advantage of the design rules is that it exposes join points that can also be used
by other AspectJ libraries as well.

Concerning correctness, when the programmer applies multiple pointcuts to
the same join point, the SM an DM libraries use aspect precedence to deal
with the composition order of their internal mechanism and of the mechanism
in a hybrid environment. For example, in the MD example the DM pointcut
for will enter first, then the SM pointcut *for* and thread reduction will happen
first than the DM reduction. The programmer does not need to deal with such
problems, however he is responsible for logically choosing where to inject the
mechanism. If the programmer injects the constructor *single* over a method and
inside this method also injects the *barrier* constructor it will result in a deadlock.
Nevertheless, programmers can define their precedence aspects and add them to
the library to deal with precedence between our libraries and other AspectJ
libraries.

3   Performance Evaluation

This section evaluates the libraries performance, against Java-based implementations using traditional PPP (i.e., non-modular). The test platform is a cluster
with two machines connected by a Gigabit Ethernet. Each machine has two E5-
2695v2 processors, each processor with 12 cores connected to a memory bank
(a NUMA with 24 physical cores per machine with 48 hyper-threading). The
cluster runs Cent OS 6.3, OpenJDK 1.8.0_20 and OpenMPI 1.8.4.

The first test uses JGF [2] Sects. 2 and 3 multi-threaded benchmarks
(JGF_MT) as the comparison base. Performance results are the speedup relative
to the JGF sequential code and also includes a JOMP[1] version. In most benchmarks the performance is comparable (Fig. 7), in some cases our SM library is
faster on others is slower. This behavior happens because performance is sensible
to many platform details. Overall our library is 1.05× slower due to overheads
introduced by aspects and the application of the design rules. The JOMP implementation is 1.2× slower and does not provide a MD implementation. The second

---

[1] The implementation of OpenMP for Java.

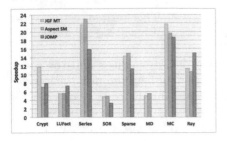

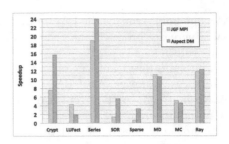

Fig. 7. JGF, JOMP vs AspectJ SM library.

Fig. 8. JGF MPI vs AspectJ DM library.

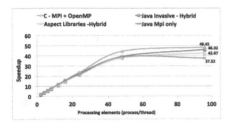

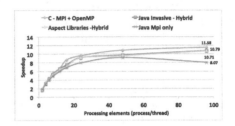

Fig. 9. MD MPI vs hybrid version.

Fig. 10. MM MPI vs hybrid version.

test uses JGF MPI benchmarks (JGF_MPI) as the base of comparison. In most benchmarks the performance of our DM library is better (Fig. 8) due to a faster implementation than the one provided by the JGF (1.4× faster).

The last two tests are a MD with a force calculation, similar to the one presented in Fig. 1 and a matrix multiplication using a tiling approach (Fig. 11). As presented in Fig. 11 our matrix multiplication instead of performing a multiplication element by element of the matrix C, uses a tiling approach based on [15]. The matrix multiplication is sub-divided into the multiplication of smaller

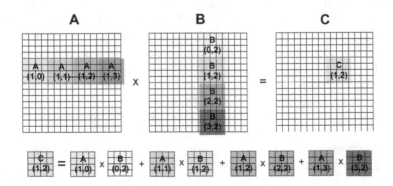

Fig. 11. Matrix multiplication with tiling.

matrices (tiling). Our algorithm was thoroughly developed to further sub-divided those smaller matrices into even smaller ones in order to fully take advantage of three levels of cache (l1, l2 and l3).

The third test evaluates the impact of composing the SM and DM libraries using two machines (24 cores each with 48 hyper-threading). The Fig. 9 compares the performance of the base Java MD (with half a million particles simulation) using only MPI processes against three hybrid versions using: (i) Java threads and MPI (non-modular); (ii) our SM and DM aspect libraries (modular); (iii) using a C version of (i) with MPI and OpenMP. The hybrid versions use one MPI process per machine, each composed by multiple threads (from 1 to 48 threads in total). The C version presents the best speedup for 96 processing elements (48×). Our version has a small overhead compared with the non-modular Java version, but both versions have performance close to the C. Using the full processing available the pure MPI version has the worst performance (37×), due to the overhead of inter-process communication. With traditional PPP, moving from this version to a hybrid version requires changes to the base program. In our approach changes are made by simply modifying the parallelism modules to be composed with the base program. The hybrid version uses a static loop scheduling among MPI process and a dynamic loop scheduling among the threads within a process. After a few tests we concluded that this strategy provides the best performance. With our approach, testing various scheduling strategies simply required a change of the pointcut in SM and DM concrete aspects. In contrast, a non-modular design requires invasive changes to the base program (e.g. Figure 1). The MD case study scales well with the number of cores. However, with the last test (Fig. 10) that evaluates the performance of a parallel matrix multiplication (using 8192 × 8192 size matrices) it does not scale so well, since it requires more communication among processes. In this test the C version also presents the best performance with 11.58× speedup for 96 processing elements closely follow up by our Java aspect version (10.71×). Finally, the performance of both Java hybrid versions are also better than pure MPI versions.

4   Related Work

Although annotations based approaches such as OpenMP and JOMP[2] [4] allow the division between domain concerns and PRC, it is restricted to the basic PRC. Sophisticated approaches required the use of explicit constructs, such as threads ids, object locks and so on. Furthermore, not only those annotations are still tangled with the base code, limiting its composability and modularity properties, but also deal only with SM PRC. Those problems are even worse with MPI libraries where only communication functions are provided (e.g., missing task distribution) which are explicitly added into the source code. Our SM and DM aspect libraries overcome those limitations by providing design rules to make the code parallel-awareness without breaking its sequential semantic. Further on, using concrete aspects as neutral zones where PRC can be expressed using

---

[2] A proposal OpenMP for Java.

a pointcut based style language. Providing an overall approach that allows to easily compose multiple PPP (e.g. SM and DM) without the need to learn two different programming languages syntaxes and fully decoupling the base code from the parallel code promoting a more cleaner and modularized approach.

Skeleton [6] frameworks, a concept proposed to encapsulate the details of a particular parallelism exploitation pattern, provide compositional proprieties, with Lithium [7] and JaSkel [8] being Java examples of such frameworks. In this kind of framework, it is necessary to create classes that will represent tasks to be done and instantiate a particular skeleton to coordinate the task execution. This approach has three main limitations: (1) the base program is polluted with scaffolding code to redirect execution the skeleton framework; (2) skeletons only encapsulate simple parallelism models (e.g., farm, pipeline ..); (3) it would be difficult to implement multiple parallelism alternatives on top of this approach, as execution issues are delegated to the skeleton framework. The last limitation can be ameliorated by the use of dependency injection (DI) [9].

In [1,11] aspect oriented programming was used to decouple PRC from domain concerns and encapsulate it in separate modules, to do so [11] used a template-based language. The work in [10] used reusable aspects to encapsulate concurrency patterns. Our work differs from these, by providing libraries (SM and DM) with competitive performance and easy to be composed (hybrids), that mimic OpenMP and MPI constructs for Java. In our approach the join point model for loops of [12] could have been used to avoid applying design rules into parallelisable loops. However, method refactoring of loops promotes independent development since the parallelisation modules depend on this explicit API. Although this approach might appears drastic at first, in reality PPP like Intel Parallel Task Library [5] follow the same strategy and some languages (e.g., R and Haskell) have higher order functions that can be seen as loop encapsulation. Exposing loops at the object interface level is the key to enable aspect modules that implement loop scheduling strategies. [7] introduced the concept of asynchronous advice, a technique to delay the execution of the code associated to a pointcut. The idea is similar to delay execution of certain blocks of code, which can also be used to introduce parallelism.

5   Conclusion

The paper presented an alternative modularisation strategy where PRC are encapsulated as aspects modules. We use the potentialities of AspectJ to support composition of multiple aspect libraries with the same base program to provide efficient hybrid solutions. Performance results show that the framework provides a competitive performance comparing with handcrafted approaches. Moreover, the hybrid versions shown to be faster than versions using only DM.

Future work will focus on the introduction of mechanisms to support parallelism using aspects for other platforms (e.g., GPUs and GRID). Furthermore, it is expected the creation of name conventions and preprocessing tools to the automatisation of the design rules.

Acknowledgement. This work has been supported by FCT - Fundação para a Ciência e Tecnologia within the Project Scope UID/CEC/00319/2013 and Search-ON2, HPC infrastructure of UMinho, NORTE-07-0162-FEDER-000086, under NSRF through ERDF.

References

1. Sobral, J.: Incrementally developing parallel applications with AspectJ. In: IPDPS 2006 (2006)
2. Smith, J., Bull, J., Obdrzlek, J.: A parallel java grande benchmark suite. In: SC 2001 (2001)
3. Kiczales, G., Hilsdale, E., Hugunin, J., Kersten, M., Palm, J., Griswold, W.G.: An overview of AspectJ. In: Knudsen, J.L. (ed.) ECOOP 2001. LNCS, vol. 2072, pp. 327–354. Springer, Heidelberg (2001). doi:10.1007/3-540-45337-7_18
4. Bull, J., Kambites, M.: JOMP an OpenMP-like interface for Java. In: JAVA 2000 ACM (2000)
5. Leijen, D., Schulte, W., Burckhardt, S.: The design of a task parallel library SIG-PLAN Not. 44, 10, pp. 227–242 (2009)
6. Cole, M.: Algorithmic skeletons: structured management of parallel computation. Pitman/MIT press, Cambridge (1989)
7. Ansaloni, D., Binder, W., Villazn, A., Moret, P.: Parallel dynamic analysis on multicores with aspect-oriented programming. In: AOSD 2010, pp. 1–12. ACM (2010)
8. Ferreira, J.F., Sobral, J.L., Proença, A.J.: JaSkel: a java skeleton-based framework for structured cluster and grid computing. In: CCGRID 2006 (2006)
9. Chiba, S., Ishikawa, R.: Aspect-oriented programming beyond dependency injection. In: Black, A.P. (ed.) ECOOP 2005. LNCS, vol. 3586, pp. 121–143. Springer, Heidelberg (2005). doi:10.1007/11531142_6
10. Cunha, C., Sobral, J., Monteiro, M.: Reusable aspect-oriented implementations of concurrency patterns and mechanisms. In: AOSD 2006, Bonn, Germany (2006)
11. Gonçalves, R., Sobral, J.: Pluggable parallelization. In: Proceedings of the 18th ACM International Symposium on HPDC 2009, Munich, Germany, pp. 11–20 (2009)
12. Harbulot, B., Gurd, J.: A join point for loops in AspectJ. In: Proceedings of the 5th International Conference on AOSD 2006, pp. 63–74. ACM (2006)
13. Laddad, R.: AspectJ in Action: Enterprise AOP with Spring Applications, 2nd edn. Manning Publications Co., Greenwich (2009)
14. Chitchyan, R., Greenwood, P., Sampaio, A., Rashid, A., Garcia, A.F., da Silva, L.F.: Semantic vs. syntactic compositions in AO requirements engineering: an empirical study. In: Sullivan, K.J. (ed.) AOSD, pp. 149–160. ACM (2009)
15. Smith, T.M., van de Geijn, R., Smelyanskiy, M., Hammond, J.R., Van Zee, F.G.: Anatomy of high-performance many-threaded matrix multiplication. In: Proceedings of the 2014 IEEE 28th IPDPS 2014, Washington, USA, pp. 1049–1059 (2014)
16. Rajan, H., Sullivan, K.J.: Classpects: unifying aspect- and object- oriented language design. In: Proceedings of the 27th ICSE, St. Louis, MO, USA, pp. 59–68 (2005)

Gaspar Data-Centric Framework

Rui Silva$^{(\boxtimes)}$ and J.L. Sobral

Centro ALGORITMI, Braga, Portugal
`ruisilva@di.uminho.pt`

Abstract. This paper presents the *Gaspar data-centric framework* to develop high performance parallel applications in Java. Our approach is based on data iterators and on the map pattern of computation. The framework provides an efficient data *Application Programming Interface(API)* that supports flexible data layout and data tiling. Data layout and tiling enable the improvement of data locality, which is essential to foster application scalability in modern multi-core systems. The paper presents the framework data-centric concepts and shows that the performance is comparable to pure Java code.

Keywords: Java · Locality optimisations · Parallel application · Data layout · Data tiling

1 Introduction

The high performance in modern computers is achieved by exploiting parallelism and accessing data efficiently. The memory hierarchy of multi-core systems is quite sophisticated whose behaviour is hard to predict. Finding the best data locality optimisations is an arduous task as it usually requires testing different approaches and parameters. The effectiveness of each optimisation may depend on the particularities of a given platform, compiler and even the application input data. Thus, programming environments should include programming constructs to implement common locality optimisations, since this is the key to the development of efficient applications for many-core systems.

The *Gaspar* framework main goal is to deliver a software platform where locality optimisations can be quickly tested. In the long term this is a path to automate locality optimisations in well-known application domains. To the authors knowledge this is the parallel programming framework that supports the most complete set of locality optimisations and the only framework that provides an infrastructure where programmers can develop application-specific locality optimisations.

The framework provides a data API that enables transparent changes to the data layout. On the other hand, the framework uses a map pattern of computation which provides an uniform mechanism to express data tiling and parallelism. Furthermore, the user can develop application-specific locality optimisations by using the provided data *API*.

© Springer International Publishing AG 2017
I. Dutra et al. (Eds.): VECPAR 2016, LNCS 10150, pp. 234–247, 2017.
DOI: 10.1007/978-3-319-61982-8_21

The next section describes the framework data *API* and the map pattern of computation and Sect. 3 provides performance evaluation. Section 4 discusses related work and Sect. 5 concludes the paper.

2   Data-Centric Framework

The framework provides two generic approaches to enable data locality improvements/tuning: (i) encapsulates the data into framework provided collections and accesses the data using iterators that hide the concrete data layout; and (ii) computation is expressed by a map & reduce pattern of computation over the framework data collections enabling changes to the order of accessing the data. These two mechanisms provide an infrastructure to implement common data locality optimisations: (i) framework provided data collections and iterators enable flexible data representations, since data is accessed using a data *API* that hides the concrete data organisation; (ii) the map & reduce pattern of computation enables changes to the order of accessing the data, since there is no explicit order in applying a function to each datum. Both mechanisms support hierarchical data organisation and processing, an essential feature to take full advantage of the memory hierarchy of parallel systems: (i) a data collection can be transparently replaced by a multi-level data representation; (ii) a map & reduce pattern of computation can be naturally decomposed into several map & reduce functions over subsets of the data. The next two subsections describe these mechanisms.

2.1   Data Application Programming Interface

The data layout is of extreme importance in modern computer architectures [6, 10]. Common data layouts for a collection of objects (or data structures) are (Fig. 1): Array of Pointers (AoP), Array of Structures (AoS) and Structure of Arrays (SoA). *SoA* layouts store the object data fields into separate arrays and are commonly used in *GPUs* to promote memory accesses coalescing [12]. *AoP* and *AoS* are more common when running applications on CPU, since they are closer to real world entities (e.g., programmers can work with data structures instead of array indexes). In certain situations it is beneficial to use hybrid layouts, where fields of a given object are stored using different layouts.

In order to achieve high performance, programmers must frequently implement data structures using rows of primitive data types (e.g., using a *SoA* layout). *Intel* suggests to use the *SoA* layout to enable auto-vectorisation on modern compilers[1]. However the *SoA* layout makes the code less legible and conflicts with modularity. To illustrate the programming complexity and the conflict with modularity this paper shows the implementation of a Molecular Dynamics simulation (MD) taken from the *JGF* [11]. One important part of this code is the computation of forces between two particles, which depends on each particle position in the space (x, y, z) (see Fig. 2).

---

[1] https://software.intel.com/sites/default/files/article/392271/
aos-to-soa-optimizations-using-iterative-closest-point-mini-app.pdf.

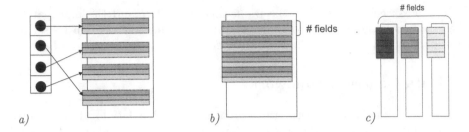

Fig. 1. Array of Pointers (AoP), Array of Structures (AoS) and Structure of Arrays (SoA) data layouts

```
//AoP layout             //SoA layout                      //Generic layout
forceParticle(...){      forceParticle(..., int id){       forceParticle(...){
  xi = p1.x;               xi = p1.x[id];                    xi = p1.getX();
  yi = p1.y;               yi = p1.y[id];                    yi = p1.getY();
  zi = p1.z;               zi = p1.z[id];                    zi = p1.getZ();
  (...)                    (...)                             (...)
}                        }                                 }
```

Fig. 2. Force computation using specific layouts and the generic layout supported by the framework.

To change from an *AoP* layout to a *SoA* or a *AoS* layout the *computeForce* method must be changed accordingly: it must receive a different kind of parameters and adapt its implementation (Fig. 2). Moreover, the *AoP* layout is the most natural implementation as the code includes a concept from application the domain: a Particle. The change from this layout to one of the others requires a rewrite of all places where the Particle class is used in the code. This was indeed required in the *JGF* MD case study used in this paper.

An alternative is to encapsulate the particle representation into a class of particles, for instance, containing tree arrays (x, y and z). That approach requires creating *Particle* objects with the sole purpose of complying to an *API* which receives *Particles* as parameters (see Fig. 2 *Generic layout*). Figure 3 illustrates this problem: it shows the creation of an object *E* (returned by method *getElement*) that will be later de-constructed (by method *setElement*). There is an object creation in order to obey a specific *API* based on objects of type *E*.

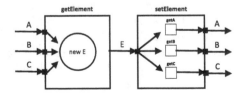

Fig. 3. Encapsulation example using generic get and set methods.

One key aspect of the framework is to support changes to data layouts without requiring changes to the *API* of methods (i.e., preserving modularity) and avoiding additional overheads. The goal is to provide a high level interface to access data while providing performance similar to low level implementations.

The framework provides a data *API* that was carefully designed to avoid runtime overheads and to not block common compiler optimisations (e.g., loop unrolling). The data *API* hides the concrete data representation making programs more portable and platform neutral.

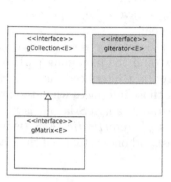 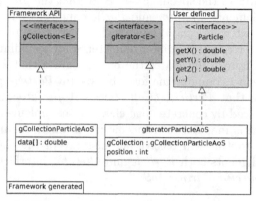

(a) *Gaspar data-centric frame-work* data API

(b) Example of framework generated classes

Fig. 4. *Gaspar data-centric framework*

The framework can generate *Java* classes that implement the data *API* for well-known types of data collections. The framework currently supports the two most common types of data structures (Fig. 4a): (i) the *gCollection* is a vector of data element; and (ii) the *gMatrix* is a set of data organised by rows and columns.

The framework data *API* is inspired by the *C++* Standard Template Library (STL), providing a similar way to iterate over data but tuned to the *Java* language. A *STL-like* approach was selected because it is widely used and it is efficiently implemented by modern *C++* compilers. The core of the data *API* is the interface *gIterator* (see Fig. 4a), which points to an element in a *gCollection*. The *gCollection* implements the methods *begin()* and *end()*, that provide iterators to the first and to a position over the last element of the collection, respectively. The iterator provides an *inc()* method to advance to the next data element and an *isLess()* method that compares two iterators. Iterators also provide the *sync()* operation which synchronises the position of two iterators. It is particularly suitable when the element position is used to identify the entity (e.g., in Matrix Multiplication) or to access its neighbours (e.g., stencil operations) on the some or on different data collections.

The framework provides a data layout generator tool that uses a *UML* data model of the application (e.g., the *Particle* specification) and generates all required interfaces and classes (Fig. 4b). Thus, it is possible to develop a program without details about the data layout and later select a collection implementation with the most appropriate data layout (e.g., *AoP*, *SoA* or *AoS*). Figure 4b provides an example from the code in the *MD* case study from the *JGF* benchmark. The programmer defines the interface of the data element (*Particle* in this case) and writes the code using the framework *API*. The framework generates the collection implementation (*gCollectionParticleAoS* and *gIteratorParticleAoS* in the specific case). Note that the *gIteratorParticleAoS* simultaneously implements *gIterator* and *Particle* interfaces, so the iterator can be used in any context where the type Particle is expected, using a Java (safe) type cast. *STL C++* requires a pointer dereference for a similar goal.

Figure 5 illustrates the base program that a programmer needs to write in order to compute the force between a *Particle p1* and the remaining particles. Note that *Particle*, *gCollection* and *gIterator* are *Java* interfaces, that will be replaced by concrete *Java* classes that implement those interfaces when the program is mapped to a specific platform. If, for instance, an *AoS* layout is used, this routine will receive one instance of the class *gIteratorParticleAoS* (which implements both *Particle* and *gIterator* interfaces) and one instance of the class *gCollectionParticleAoS*.

```
// the same method for all data layouts
void forceParticle(Particle p1, gCollection<Particle> particleSet) {

    // get coordinates of particle p1
    xi = p1.getx();
    yi = p1.gety();
    zi = p1.getz();

    // iterate over particleSet
    gIterator<Particle> p2 = particleSet.begin()
    for( ; p2.isLess(particleSet.end()); p2.inc()) {

        // compute distance
        xx = xi - ((Particle) p2).getx();
        yy = yi - ((Particle) p2).gety();
        zz = zi - ((Particle) p2).getz();
        (...)
    }
}
```

Fig. 5. Example of the usage of the data API

To understand the reason why there is no runtime overhead, Fig. 4b also shows the implementation of the *gIteratorParticleAoS* and *gCollectionParticleAoS* classes: the former contains an integer, which is an index to an array of doubles (the latter). In this case, the *JIT* is able to generate an implementation of the *forceParticle* method from Fig. 5 tuned for classes *gIteratorParticleAoS* and *gCollectionParticleAoS*. Therefore, all iterators are replaced by an integer index into an array of doubles, which is essentially a raw *AoS* implementation. Thus, the *JIT* can generate an executable equivalent to an AoS-based hand-coded. Note that *STL* iterators also rely on compiler optimisations, namely, method inlining. The *JIT* compiler additionally relies on escape analysis to avoid the creation of objects that are instances of *gIterators*.

2.2   Map Pattern

One most common data locality optimisation is the use data tiles to improve temporal locality [2]. Traditionally this optimisation requires the introduction of additional loops to implement operations over data tiles, where some base kernel is applied on data tiles in cache. It may be also beneficial to implement multiple levels of tiling to address multiple levels of cache. Each level of tiling will require additional loops.

The framework uses a map pattern of computation that is able to express both tiling-based locality improvements and parallel computations over *gCollections* of data. The map pattern is based on a *map method* which applies a given function to a *gCollection*, by iterating over the collection elements using *gIterators*.

The map pattern divides a *gCollection* into multiple collections using a *splitFunction*, applies a *mapMethod* to each sub-collection and invokes a *reduceFunction* to join the generated/processed sub-collections. The map pattern can generate additional data copies that can lead to inefficient implementations. The framework data *API* enables two mechanisms to improve performance of map operators: virtual collections and lazy copying. The framework offers several splitter and reducer functions in order to use these optimisations. One implementation creates a physical copy of the data, which can improve the spatial locality (an operation also known as packing). This implementation has the option to do the packing of all sub-collections when the split function is called or when the each sub-collection is accessed (i.e., lazy packing). Virtual collections avoid performing additional data copies by creating collections and iterators that are virtual views of the original data.

The map is implemented by an high-order function introduced in *Java 8* (i.e., a function that accepts pointers to methods). The map function has the following interface: *MAP(splitFunc, mapMethod, reduceFunc, gCollection)*.

Figure 7 illustrates the map pattern applied to the force method of our illustrative case (see the force method defined in Fig. 6). The tiling optimisation on this case requires nesting of two map patterns, where the inner loop depends the current iteration of the outer loop. The outer map (line 3) divides the collection *c1* applying the *splitc1* function (lines 9–20), and calls the inner map (line 4–5) for all sub-collections. The inner map applies the same procedure for collection *c2*, but calls the original function *force*. Finally, the outer map applies *reducec1* function (line 5) to the processed sub-collections and returns the result. The

```
void force(gCollection<Particle> c1, gCollection<Particle> c2) {
   gIterator<Particle> p1 =  c1.begin();
   for(; p1.isLess(c1.end()); p1.inc())
     forceParticle( (Particle) p1, c2);
}
```

Fig. 6. Force computation for all-particles

```
1  void forcemap(gCollection<Particle> c1, gCollection<Particle> c2) {
2  Object parameters = new Parameters( c1, c2);
3  parameters = gCollection.map(splitc1,
4    (Object m) -> gCollection.map(
5            splitc2, force, reducec2, m),
6      reducec1, parameters)
7  }
8
9  gCollection<Object> splitc1(Object obj){
10 int threads = Integer.getInteger("threads", 2);
11 Parameters parameters = (Parameters) obj;
12 Parameters auxret[] = new Parameters[threads];
13 gCollection value= new parameter.p1.split(threads);
14 for(int i=0; i < auxret.length; i++){
15   Parameters newparameters = new Parameters(value.get(i),parameters.p2);
16     auxret[i] = newparameters;
17   }
18   gCollectionObject ret = new gCollectionObject<Parameters>(auxret);
19   return ret;
20 }
21
22 Object reducec1(Object value, Object obj){
23   return ((gCollection) obj).reduce();
24 }
```

Fig. 7. Force computation with tiling optimization

splitc1, *splitc2*, *reducec1* and *reducec2* are defined using the framework built-in split and reduce operations.

3   Evaluation

This section presents several case studies that evaluate the framework performance. The first subsection analyses the code generated by the *JIT*, comparing versions generated with the *Gaspar framework* and in plain *Java*. This uses a simple benchmark that sums all values in a collection of doubles. The second subsection presents performance using two case studies: Matrix Multiplication and Molecular Dynamic Simulation. The benchmarks were executed on a machine with two 12-core Xeon E5-2695v2, running the OpenJDK 1.8.0_25. In all tests use the following flags to optimize the Java Virtual Machine: *-XX:LoopUnrollLimit=100*; *-XX:ObjectAlignmentIn Bytes=32*; *-XX:+UseNUMA*; *-XX:+UseCompressedOops*. Presented values are a median of 10 runs.

3.1   Low Level Benchmark

This benchmark compares the performance of *gCollection* and *Java* implementations. The first type of *Java* implementation uses *Java Collections* (*ArrayList<Double>*) of type Double (*AoP*, see Sect. 2.1). There are three different versions: *fAoP* uses *For Each* to access data; *iAoP* access data with *get(i)*; *sAoP* uses *Java 8* streams. The second type of Java implementations uses arrays (the arrays contains data, not pointers to data): *iSoA*; *fSoA*; *sSoA*. The Fig. 8 shows *Java Code* for all *AoP* layout implementations, note that the framework uses the same code for all layouts (e.g., AoP and SoA).

```
//Sum (fAoP)                      //Sum without iterators (iAoP)
   for (Double value: aopy) {        for (int i=0;
                                   i < aopy.size(); i++) {
      result += value;                 result += aopy.get(i);
   }                                 }

//Sum with stream (sAoP)          //Framework Sum (gAoP or gSoA)
   result =  aopy.stream().           gIterator it = gaopy.begin();
 mapToDouble(Double::doubleValue)  for (; it.isless(gaopy.end());
 .sum();                           it.inc()) {
                                      result += ((gDouble) it).
                                   getValue();
                                      }
```

Fig. 8. Java code of different implementations for AoP layout

The benchmark analyses two array sizes (Fig. 9): $6.4 \times 10^5 (\sim 5\,\text{MB})$ and $5.12 \times 10^7 (\sim 390\,\text{MB})$. In the small size the data fits into the level 3 cache, while in the other size the data is accessed from memory. This causes an increase in the cycles required to perform each instruction (CPI increase). *Gaspar* implementations (*gAoP* and *gSoA*) are able to deliver a performance equal to the best Java implementation (iAoP and iSoA cycles per element). The results for small and big data size follow the same tendency, thus, Fig. 10 focus on results for the larger size.

The Fig. 10a shows that *SoA* layouts have better performance than *AoP* layouts (6 times better for *fSoA*, *iSoA* and *gSoA*). *SoA* layouts access the data directly, while *iAoP* access the data through a pointer indirection (see code in Fig. 11). This explains the lower number of instructions executed in *SoA* layouts. In *SoA*, data is accessed more efficiently, since data is aligned in memory, which results in less cache misses (Fig. 10b) due to better spatial locality.

Using the *For Each* in layout *AoP* has a small impact on performance (see *fAoP*). This impact results from the increase in number of instructions, because *For each* in Fig. 9 implements one *Cast Check* per element while the *iAoP* performs one *Cast Check* per 8 elements (due to more efficient loop unrolling). This additional overhead does not exists in the fSoA layout.

The use of streams of *Java 8* shows significant decrease in performance (∼0.57 in *fAoP*, ∼0.25 in the *fSoA*) since it uses Kahan [5] algorithm for reducing truncation errors.

The *iSoA* needs 21 instructions to process 16 elements (unroll 16 times), but in the framework (*gSoA*) 23 instruction are generated to process the same elements. One additional instruction writes the position of the iterator in memory. The other is to due bad optimization heuristics (see the assembly in Fig. 12, uses one more register).

Although there are no performance differences in *AoP* layout, *gAoP* has less instructions since the compiler was able to remove instructions of *Java Exception* for control the cast to *Double*.

So for this case our framework has no impact on performance and is able to change data layout without development cost (it is possible change layout and to enable performance effortlessly).

3.2   High Level Benchmark

The first case study is a Matrix Multiplication(MM), we compare code in plain *Java* with use of the framework. The kernel used is the same in both implementations, as well as the tiling optimisation. The Framework with tiling is based on the map operator described in Sect. 2.2. Figure 13a shows the relative performance of tiling and lazy packing. The data *API* introduces a small overhead. In the plain Java implementation, tiling improves the performance, but with a traditional map pattern implementation advantage of tilling is lost due to copying overhead. The introduction of lazy packing provides a performance comparable to a plain Java implementation. The figure also presents the execution time for

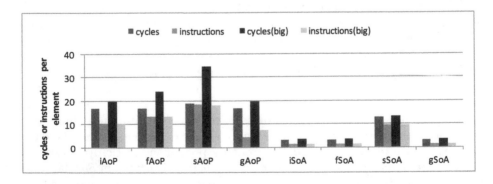

Fig. 9. *Sum* number of instructions and cycles per element

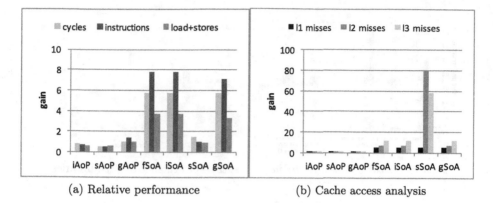

(a) Relative performance   (b) Cache access analysis

Fig. 10. Sum relative performance to fAoP

```
//Assembler for iAoP load and sum element   //Assembler for iSoA load and sum element
mov    0x10(%rbx,%r8,4),%r11d               vaddsd 0x10(%r11,%rbx,8),%xmm0,%xmm0
vaddsd 0x10(%r10),%xmm0,%xmm0
```

Fig. 11. Differences from *iAoP* to *iSoA* (load and add element)

```
//Assembler for iSoA sum                    //Assembler for gSoA sum
...                                         ...
vaddsd 0x80(%r11,%r9,8),%xmm0,%xmm0         vaddsd 0x80(%r10,%r8,8),%xmm0,%xmm1
vaddsd 0x88(%r11,%r9,8),%xmm0,%xmm0         vaddsd 0x88(%r10,%r8,8),%xmm1,%xmm0

vmovsd %xmm0,0x70(%r10)                     vmovsd %xmm1,0x70(%rbp)
...                                         vmovsd %xmm0,0x70(%rbp)
                                            ...
```

Fig. 12. Compiler overhead problem in *gSoA*

lazy packing of all matrices (A, B and C) and only for matrix C, which was tested by simply changing split functions.

Figure 13b presents the relative performance of this version against other well-known pure Java implementations (the reference implementation is the *JBLAS* implementation). The framework provides the best pure Java implementation and up to 0.95 times the performance of the *JBLAS* implementation (*JBLAS* provides 0.70 times of the peak performance on this machine).

Tuning performance is a hard task and finding the best matrix implementation requires experimentation, since there are three nested loops in the MM. In the framework, the experimentation can be quickly performed by adding a nested map. Figure 13c illustrates the relative performance using the *ti, tj, tk* order as reference, by changing the map nesting order. The best order also depends on the input size. In the framework this can be addressed by using different map nests for each input size.

A parallel version of the *MM* is developed by replacing the map implementing the tiling with a parallel map. Figure 13d presents the speed-up obtained with

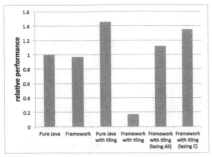

(a) *Java* vs *Gaspar data-centric frame-work*

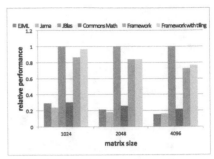

(b) Performance of *Java* libraries

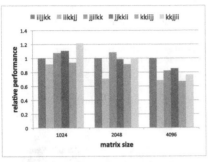

(c) Differents tile orders

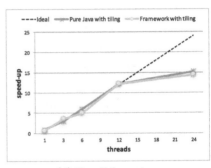

(d) Parallel versions

Fig. 13. Matrix Multiplication benchmark

this implementation and the comparison with an equivalent implementation in plain *Java*. The performance of both implementations is very close and both scale linearly up to 12 processors. However, for 24 threads there is a performance penalty in both versions, caused by load unbalance (some threads process one more block than the others) and caused by the *NUMA* architecture.

The MD benchmark from the *JGF* (using an *AoP* layout) was implemented in the framework and tested with different data layouts (gXoX versions). The speed-up to the sequential *SoA* version (the more efficient) of each version is presented in Fig. 14b. The *AoP* layout scales poorly due to the lack of data locality and because its sequential version is the slowest. The performance of the framework *AoP* implementation is similar to one of *JGF*. The *SoA* version provides the best performance.

The data layout can interplay with tiling. The framework provides a flexible mechanism to develop custom tiling approaches by implementing case-specific split/reduce functions for the map. A custom tiling approach is required for *MD* benchmark since there is a nested loop, where the inner loop depends on the current iteration of the outer loop. Figure 14a presents performance by composing the different data layouts of MD with tiling.

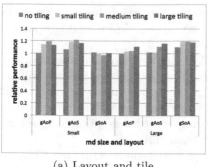
(a) Layout and tile

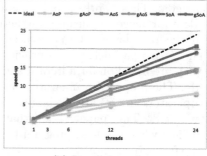
(b) Parallel versions

Fig. 14. Molecular dynamic benchmark

For a small particle set, the *AoS* is the best layout and tiling improves the performance. Performance of *AoP* and *AoS* are very close for medium tiles. However, tiling does not improve the *SoA* version, which presents the worst performance. On the larger size, the *SoA* is the best version and, in this case, it benefits from applying tiling. In parallel versions, for all layouts in the *Gaspar framework* the performance is similar to versions in plain *Java*. The *SoA* layout obtains the best performance. In the plain *Java* the best speed-up is 20.8. The framework has a small loss in performance, the speed-up is 19.1.

4 Discussion and Related Work

There is a number of techniques to automatically improve locality by changing the *Java Virtual Machine(JVM)*. Hirzel et al. [4] evaluated a technique based on sorting objects during garbage copying, which places objects in consecutive memory addresses to improve spatial locality. This technique still maintains the *AoP*. Wimmer et al. [14] propose an improvement to the *JVM* to automatically inline object fields by placing the parent and children objects in consecutive memory places and by replacing memory accesses by address arithmetic. Nie et al. [9] propose the Java vectorisation interface to explicitly expose data parallelism in programs enabling explicit vectorisation. These works require changes to the *JVM* implementation and there is no known system that supports data tiling.

OpenACC and *Mint* [13] are two programming frameworks that provide *OpenMP* like directives to support the loop tiling by a specific loop clause. The *Gaspar data-centric framework* provides a more flexible approach, for instance, it is easily change the tiling order or change the data layout.

There are several alternative implementations of *Java* generics that avoid some of the *Java* overhead. The High Performance Primitive Collections[2] provides template generated collections for primitive data types. Trove[3] shares a

---

[2] http://labs.carrotsearch.com/hppc.html.
[3] http://trove.starlight-systems.com.

similar goal. These works avoid generics overhead on primitive data types, but they do not remove overheads on structured data types, thus they do not support alternative data layouts for structured types.

The *Java 8* parallel streams provide an API that resembles to map operators, but they are based on Java iterators and do not support data tiles and different layouts. The proposed framework uses a more sequential-like way of expressing map operators and is part of a larger effort to provide *OpenMP-like* programming in Java [7,8]. This paper shows that the map pattern is suitable to express parallelism when the base program benefits from tiling.

The framework data *API* is similar to Standard Template Library(STL), but in *STL* there is a difference between a pointer (iterator) and the element pointed to. Thus, to access an element, pointer dereference must be used so it is not possible to automatically transform an iterator to the element pointed to. This will make it more difficult to encapsulate the data layout, without introducing a concept similar to interfaces in *Java*.

Other approach is to hide the distribution and parallelism concepts in skeletons. *STAPL* [3,15] and *FastFlow* [1] provide skeletons to simplify the code and improve performance. In *STAPL*, the skeletons express data distribution and parallelism, it is and based in *STL* iterators. Although, both frameworks do not support multiple data layout.

The map pattern of computation with iterators provides a safer way of iterating over data than using the traditional loop-based approach, since, in a map, the loop range is implicitly derived from the collection size. This avoids many potential errors of a loop-based approach, specially when multiple levels of tiling are required.

5   Conclusion

This paper presented a *Gaspar data-centric framework* and how the proposed data *API* efficiently supports multiple data layouts and tiling. The performance results show that the framework can provide implementations that compete with pure *Java* implementations. Thus, the framework provides improved trade-off between programmability and performance. The framework data *API* was also designed to make it easy to introduce locality improvements in loops that iterate over data in a collection. The provided infra-structure makes it easy to tests different data layouts and tiling, as well as to develop case-specific locality optimisations.

In future, the tool will provide a performance analyser helping the user to select data locality improvements. On the other side it will be included support to more computing platforms (e.g., distribute memory).

Acknowledgements. This work has been supported by *FCT - Fundação para a Ciência e Tecnologia* within the Project Scope *UID/CEC/00319/2013* and *Search-ON2, HPC infrastructure of UMinho, NORTE-07-0162-FEDER-000086, under NSRF through ERDF*.

References

1. Aldinucci, M., Danelutto, M., Kilpatrick, P., Meneghin, M., Torquati, M.: Accelerating code on multi-cores with FastFlow. In: Jeannot, E., Namyst, R., Roman, J. (eds.) Euro-Par 2011. LNCS, vol. 6853, pp. 170–181. Springer, Heidelberg (2011). doi:10.1007/978-3-642-23397-5_17

2. Anderson, J.M., Lam, M.S.: Global optimizations for parallelism and locality on scalable parallel machines. In: Proceedings of the ACM SIGPLAN 1993 Conference on Programming Language Design and Implementation. PLDI 1993, pp. 112–125. ACM, New York (1993)

3. Buss, A., Papadopoulos, I., Pearce, O., Smith, T., Tanase, G., Thomas, N., Xu, X., Bianco, M., Amato, N.M., Rauchwerger, L., et al.: Stapl: standard template adaptive parallel library. In: Proceedings of the 3rd Annual Haifa Experimental Systems Conference, p. 14. ACM (2010)

4. Hirzel, M.: Data layouts for object-oriented programs. SIGMETRICS Perform. Eval. Rev. 35(1), 265–276 (2007)

5. Kahan, W.: Pracniques: further remarks on reducing truncation errors. Commun. ACM 8(1), 40 (14965). http://doi.acm.org/10.1145/363707.363723

6. Majeti, D., Barik, R., Zhao, J., Grossman, M., Sarkar, V.: Compiler-driven data layout transformation for heterogeneous platforms. In: Mey, D., et al. (eds.) Euro-Par 2013. LNCS, vol. 8374, pp. 188–197. Springer, Heidelberg (2014). doi:10.1007/978-3-642-54420-0_19

7. Medeiros, B., Silva, R., Sobral, J.: Gaspar: a compositional aspect-oriented approach for cluster applications. In: Concurrency and Computation: Practice and Experience (2015)

8. Medeiros, B., Sobral, J.L.: Aomplib: An aspect library for large-scale multi-core parallel programming. In: 2013 42nd International Conference on Parallel Processing (ICPP), pp. 270–279. IEEE (2013)

9. Nie, J., Cheng, B., Li, S., Wang, L., Li, X.-F.: Vectorization for Java. In: Ding, C., Shao, Z., Zheng, R. (eds.) NPC 2010. LNCS, vol. 6289, pp. 3–17. Springer, Heidelberg (2010). doi:10.1007/978-3-642-15672-4_3

10. Sharma, K., Karlin, I., Keasler, J., McGraw, J.R., Sarkar, V.: User-specified and automatic data layout selection for portable performance. Rice University, Houston, Texas, USA, Technical Report TR13-03 (2013)

11. Smith, L.A., Bull, J.M., Obdržálek, J.: A parallel Java grande benchmark suite. In: Proceedings of the 2001 ACM/IEEE Conference on Supercomputing. SC 2001, pp. 8–8. ACM, New York (2001)

12. Sung, I.J., Liu, G.D., Hwu, W.M.W.: Dl: A data layout transformation system for heterogeneous computing. In: Innovative Parallel Computing (InPar), pp. 1–11. IEEE (2012)

13. Unat, D., Cai, X., Baden, S.B.: Mint: realizing cuda performance in 3D stencil methods with annotated C. In: Proceedings of the International Conference on Supercomputing, pp. 214–224. ACM (2011)

14. Wimmer, C., Mössenböck, H.: Automatic array inlining in Java virtual machines. In: Proceedings of the 6th Annual IEEE/ACM International Symposium on Code Generation and Optimization, pp. 14–23. ACM (2008)

15. Zandifar, M., Thomas, N., Amato, N.M., Rauchwerger, L.: The STAPL skeleton framework. In: Brodman, J., Tu, P. (eds.) LCPC 2014. LNCS, vol. 8967, pp. 176–190. Springer, Cham (2015). doi:10.1007/978-3-319-17473-0_12

A Parallel and Resilient Frontend for High Performance Validation Suites

Julien Adam[1][✉] and Marc Pérache[2]

[1] Paratools SAS, Bruyères-le-Châtel, France
adamj@paratools.com
[2] CEA, DAM, DIF, 91297 Arpajon, France
marc.perache@cea.fr

Abstract. In any well-structured software project, a necessary step consists in validating results relatively to functional expectations. However, in the high-performance computing (HPC) context, this process can become cumbersome due to specific constraints such as scalability and/or specific job launchers. In this paper we present an original validation front-end taking advantage of HPC resources for HPC workloads. By adding an abstraction level between users and the batch manager, our tool *JCHRONOSS*, drastically reduces test-suite running time, while taking advantage of distributed resources available to HPC developers. We will first introduce validation work-flow challenges before presenting the architecture of our tool and its contribution to HPC validation suites. Eventually, we present results from real test-cases, demonstrating effective speed-up up to 25x compared to sequential validation time – paving the way to more thorough validation of HPC applications.

Keywords: Validation · Test-suite · HPC · Scheduling · Fault-tolerance · Parallel · Software quality

1 Introduction

In the constantly evolving landscape of parallel supercomputers, HPC applications must be updated to take advantage of the underlying architectures. In such a context, validating parallel software features can be a real challenge. Non-regression bases (NRB) can play an important role in such transitional process, constantly validating results relative to expectations – matching each features with dedicated tests. However, for larger projects, the growth of the non-regression base can become troublesome, particularly if validation system is not robust enough. A recent project with very large non-regression bases took up to several days to run and involved thousands of tests. In such a context, modifications could take up to one week to be validated, making test results more complex to analyze and impacting development reactivity. Current testing frameworks do not provide a scalable way to meet the growing validation demands of large sofware efforts.

Our goal is to simplify the continuous validation of parallel HPC applications, allowing HPC developers to constantly monitor their software quality in

© Springer International Publishing AG 2017
I. Dutra et al. (Eds.): VECPAR 2016, LNCS 10150, pp. 248–255, 2017.
DOI: 10.1007/978-3-319-61982-8_22

an efficient manner. In this paper, we present a highly modular testing framework, called *JCHRONOSS*, that provides a convenient and consistent abstraction layer between a parallel validation suite and a given batch-manager. This tool is intended to be scalable on most HPC architecture, with dynamic scheduling and resilient execution. As we will show, JCHRONOSS has been built in a generic manner, without constraining the target execution model in order to meet the requirements of any developer, conveniently replacing the commodity test-scripts encountered in some projects. The purpose is to optimize the continuous integration process by providing a quick and reliable feedback on software quality during the development process. JCHRONOSS is built in the context of existing integration testing utilities, thereby enhancing validation work-flows in an HPC context, while allowing the user to rely on standard components.

This paper is organized as follows: Sect. 2 describes related work, discussing the use and limitations of non-regression bases in HPC context. Section 3.1 shortly presents JCHRONOSS's architecture. Then, Sect. 3 details JCHRONOSS's contributions to continuous integration in HPC context and Sect. 4 evaluates JCHRONOSS in different configurations relatively to a real use-case. Finally, Sect. 5 describes open issues and future work.

2   Related Work

The main focus of JCHRONOSS is to run tests in an optimized way. This process involves two main components that we have to compare with existing work: (1) schedulers and (2) test-frameworks.

Schedulers. Resource scheduling has been widely studied for years and a large number of tools already covers the subject particularly in HPC context. For example, a tool like YARN [8] from the Apache Hadoop framework is a powerful scheduler, able to distribute multiple applications over thousands of resources, such as those used for MapReduce [5] computations. Borg [9] from Google can distribute applications over multiple clusters, each composed of thousands of nodes, with a goal of supporting a huge number of requests per second. In the HPC context, job managers such as SLURM [10] are deployed over a cluster to efficiently manage resource allocation. Such schedulers generally need to be deployed at the system level in order to expose computing resources. On the contrary, JCHRONOSS is running in user space, processing a test-suite meta-description and generating calls to such job-manager in a more efficient manner. Indeed, running a test-suite in parallel requires more than simply submitting executions to an existing batch manager, as we will further detail.

Test Frameworks. As testing is a key process to ensure software quality, there are a wide range of tools and solutions. Most solutions are focused on ease of use, especially when dealing with automatic generation and configuration aspect. CMake [6] and Autotools [4] are two main project builder, able to handle the configuration and generations of test suites in a convenient way through macros.

Some continuous integration platform like Jenkins [3], Travis [2] or CruiseControl [1] are designed to create integrated test environments gathering several key components in the same interface (such as version control systems and ticket trackers) However, these solutions were not developed for HPC, as they are not able to conveniently express the execution of their workload in parallel, this burden being left to the end-user. Developers are then forced to develop their own validation script, tailored to a given test environment. JCHRONOSS proposes to avoid this redundant effort thanks to a simple XML formatted input driving a parallel execution from user-defined templates (batch-manager agnostic), without sacrificing portability. Our tool is not a job scheduler by itself, it is designed to be run by a user to generate from an XML meta-model an optimized stream of requests to an existing batch-manager (the one installed on the machine).

3   Contribution

In this section, we present the three main contributions of our tool. First, we detail JCHRONOSS's architecture and its main components. Then, we explain how tests are scheduled over a supercomputer. Eventually, we describe the fault-tolerance mechanism. These contributions allow JCHRONOSS to use a surface-based scheduler with resiliency to run tests in parallel and optimize validation time.

3.1   Global Model

JCHRONOSS is designed for ease of use and interoperability. It loads a standard validated XML input and produces a standard JUnit formated output compliant with common continuous integration platforms. As depicted in Fig. 1, the master-worker architecture is based on two independent layers doing mostly the same processing. In order to keep resources as busy as possible, layers share the same algorithm following a "greedy" approach. Jobs are scattered in sub-pools assigned to *workers*.

Workers are responsible for executing individual sub-pools. Sub-pool resources are subtracted from a global resource allocation counter. Then, when there are no resources left, the master stops creating workers. Upon completion, results are merged in a *post-run* list gathering completed tests' results – process repeated until test-suite completion. The only difference between master and worker is their scope. The master is responsible

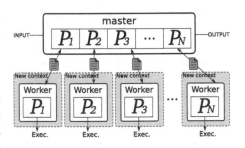

Fig. 1. Master/Worker Architecture

for the global validation system whereas a worker manages a subset of tests, effectively running them over the system.

3.2 Job Ordering

Making requests to the job manager is as important as the scheduling itself. In the context of overloaded supercomputers, the more requests are made by a user, the harder it is for the job manager to satisfy them. Generally, allocation grants are based on multiple criteria. This is why requesting 2 nodes twice is not always equivalent to a 4 node request. Allocation rate depends on current cluster load, past requests, quotas, and the number of resources. Given these constraints, the most basic test runner would make a request for individual tests. This can seriously degrade user priority, making future allocation attempts longer.

JCHRONOSS offers a way to gather jobs depending on deterministic criteria, such as number of resources. This way, if a test requests four nodes to run, JCHRONOSS will attempt to create a worker with multiple jobs requesting the same number of nodes, allocating the node configuration only once. This follows a very simple principle: if the allocation is created according to type and number of required resources, then jobs sharing similar requirements can be dependent on the same allocation. By gathering jobs with the same requirements in the same allocation, this policy tries to limit the number of resource requests, leading to larger workers (more jobs per allocation) and lowering global allocation overhead. However, as such contexts ask for more resources, the batch-manager can be a little longer to fulfill the request. But, if the batch-manager does not penalize allocation following a linear allocation time formula like $f(x) = ax$ (which is generally the case), this algorithm will always be preferable for this kind of configuration. This approach is less stressful, and best suited for homogeneous validation suites. Indeed, with imbalanced job pools, one worker will have to process more tests than the others, eventually leading to a parallelism loss.

Another approach can be considered to take advantage of a higher level of parallelism. Another solution consists in running validation suite depending on available resources instead of test requirements. The strategy evenly divides resources among workers. Then, jobs are scheduled using a two-dimensional houristic over both resources and time, the purpose being to fill each parallel subset as much as possible. Jobs are first sorted by resource requirements and then by decreasing estimated time. Thanks to this ordering, larger jobs are scheduled first, using a classical greedy scheduling heuristic. This way, JCHRONOSS can guarantee an efficient use of available resources at any time. Ideally, efficient scheduling requires a prior knowledge of individual test duration in order to correctly apply the "surface" scheduling heuristic. However, if not provided, or at least bounded by individual job timeout, JCHRONOSS approximates job duration as the mean of previous duration.

This algorithm is the most efficient for non-homogeneous test-suites in terms of job manager requests as it allocates large subset and tries to fill them – maximizing resource efficiency. However, if the batch-manager policy is resource-based, allocating large buckets can lead to very long allocation time, leading to poor performance. Nonetheless, we observed that in most cases, the best-fit policy is a good trade-off between efficiency and execution time.

3.3   Fault Tolerance

Depending on code coverage, validation suites can take a lot of time, ranging from a few minutes up to several days. However, HPC environments are not fully reliable with, for example, failing nodes, batch-manager and timeouts – possibly impacting running jobs. JCHRONOSS has been designed to be fault tolerant. It supposes that any layer can crash. If a worker is interrupted, the master considers all jobs as not run and reschedules them, making our approach completely resilient to failing workers. Indeed, a new worker will be created to replace the failing one and the tests will be rescheduled. Therefore, losing a worker has no effect on validation's coverage. The case where the master instance is interrupted is more problematic as job results are only merged at the end of the test-suite. Consequently, a crash prior to this point would lead to a complete loss of master's state. In order to circumvent this limitation, we implemented an asynchronous check-pointing mechanism which consists in storing current job states in a file as the workers are running. Thanks to this approach, a validation can be restarted from the last coherent checkpoint, even if the master instance failed, providing a complete fault-tolerance support.

Checkpoint Time. A checkpoint is initiated when the master expects a worker to end to maximize the overlapping. It consists in storing current jobs' state and their configuration. Workers do not need to be checkpointed, they will be recreated upon restart, scheduling remaining jobs. Our backup consists of a single JSON formatted file stored in JCHRONOSS's build directory, alongside other temporary files. JSON format is flexible and easy to manipulate inside JCHRONOSS, however, for now, the JSON file is not compressed and can lead to both IO and parsing overhead depending on validation suite size. We are considering the use of a binary JSON (BSON) to optimize this process.

Restart Time. After an interruption, JCHRONOSS can be restarted from the backup file. To do so, current configuration is ignored and previous one is reloaded. Then, job manager's state is restored from the backup JSON file. Finally, validation can restart seamlessly. In order to save disk space, following backup files replace previous ones. Therefore, the most recent backup is always kept and calling the same command line over again in case of failure allows the completion of an incomplete test-suite thanks to our fault-tolerance mechanism.

Overhead. We plan to make a deep evaluation of fault-tolerance mechanism overhead. For now, our experiments show that it takes 1 second per worker to back up 10,000 tests and the global overhead does not exceed 1.2%. Clearly, the number of tests can be different and the number of workers can noticeably vary depending on the user's configuration. By trying to checkpoint only validation state and not JCHRONOSS itself, we significantly decrease implied backup overhead. It is important to say that the major part of this overhead is recovered by workers instance currently running. However, this mechanism can become really costly with an important number of workers, this increasing checkpoint time, not completely recovered by shorter workers.

4 Experimental Results

JCHRONOSS has been developed for and is being used on a daily basis as MPC [7] validation system to manage a test base of forty thousand jobs, test-suite likely to be executed on several supercomputers, involving different environments for portability tests. JCHRONOSS's goal is to speedup validation processes without sacrificing their portability between machines. In this purpose, the important variability between HPC machines had to be taken into account. Indeed, as aforementioned several parameters affect scheduling such as current user priority and specific latency due to cluster load. Moreover, as the machine load is highly variable, we cannot predict allocation overhead. Then, two successive JCHRONOSS runs, with similar parameters might not lead to the same result. We were careful to present tests with similar configurations while mitigating these random effects. These benchmarks were performed on two different supercomputers. First the Curie supercomputer, operated in the TGCC, the french CEA Very Large Computing Center, which is heavily loaded by multiple users, leading to long waiting queues. The batch manager, based on SLURM is configured with user priorities. Second supercomputer is a 111 nodes × 8 cores prototype, with fewer users and a flexible batch manager. Comparisons will be made between these two environments, respectively with and without priority based algorithms applied at batch manager level. The NRB used here is a suite of 39,366 jobs with fixed execution times to allow policies comparison over multiple runs while minimizing measurement noise. These configurations have been run with the same subset of available resources, allowed to perform tests on 48 nodes. We compare policies in terms of elapsed time on each of these supercomputer. These comparisons will be made alongside CTest performance with the same set of tests. The Fig. 2 depicts these results.

Complete validation suites were run on each of these machines with different policies in order to compare batch manager configuration effects. The fixed number of resources is set to 4. Vertical axis represents the number of hours elapsed in the run. CTest results have been run sequentially (ctest -j4) to be able to compare with JCHRONOSS. Indeed, the -j option allows tests to be run in parallel without discrimination, implying job over-submissions and causing the user to violate the QoS

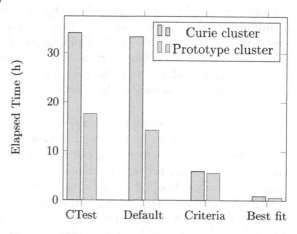

Fig. 2. Policies efficiency comparison between two supercomputer job managers.

policy and account to be blocked if the value is too high. Each job keeps the same execution time in each execution We consider that machine load variation did not impact test-suite duration between policies, Curie being loaded and our test cluster almost empty. These results illustrate the need to carefully choose scheduling policies according to cluster, allocation overhead being highly depending on batch manager. Default sequential policy clearly shows its limits, providing no performance gains on the test-cluster and leading to an important penalty on Curie. More importantly, allocation overhead even led to poor performance relatively to a sequential execution. Default policy created around 40,000 new allocation requests, each of them associated with a resource allocation, explaining the overhead observed on the loaded cluster. This policy roughly applies the same methodology as other test-runner tools, as depicted by the sequential CTest performance results.

Our criteria-based policy shows a non-negligible time reduction with a 2.5 speedup. Packing jobs relatively to resources seems to be a good alternative to sequential execution. Indeed, considering of N jobs, this solution can save up to $N - 1$ new allocations if they are all using the same number of resources.

Eventually, best fit algorithm shows the best speed-up of 25, independently from the underlying batch manager. The optimizations made by this policy, spreading jobs among resources in order to save time have proven to be effective. More importantly, this approach seems to be less sensitive to batch-manager policy, making it more suitable for portability. Best fit is then both the fastest and the most portable policy – reason why it is the default one in JCHRONOSS.

5   Future Work

Optimize time to result. Currently, all tests defined by the user must be performed before publishing the results. Therefore, it can happen that the whole test suite has to be completed before the user is able to consult the results, including intermediate ones. In order to make time to result shorter, a deamon server, provided as a JCHRONOSS plugin and running globally on interactive nodes, could interact directly with worker instances, periodically collecting job results and making data accessible from a client browser. To reduce the number of deamons, a single server would handle multiple JCHRONOSS instances.

Becoming a complete end-to-end validation tool. For now, existing validation processes would have to be rewritten in order to generate a suitable input for JCHRONOSS. We suggest making our tool compliant with upstream and downstream tools, avoiding test specifications rewriting. JCHRONOSS should include a job generator module, which could take data from existing build systems like CMake or Autotools. Dealing with the output, JCHRONOSS generates it in standard JUnit XML format. However, some other formats could be more suitable for post-processing. A generic output generation module would bring more flexibility to the end-user. Our idea is to gather in one single tool all validation steps from the build system to the result mining platform, leading to an end-to-end validation tool.

6   Conclusion

JCHRONOSS is a parallel and resilient frontend for high-performance validation suites that run distributed tests in parallel in order to reduce time to result.
Beyond just taking advantage of parallel computing resources, JCHRONOSS
looks for optimal trade-offs between efficiency and duration. Its multiple scheduling policies are suitable for most use cases, allowing JCHRONOSS to be an innovative agile tool designed for HPC workloads. JCHRONOSS can be adapted to
various execution environments and is compatible with existing validation tools
such as Jenkins and BuildBot. We demonstrated validation speedup up to $25\times$
on an actual use case of $\approx 40,000$ tests, clearly showing the advantage of our approach. JCHRONOSS is then a convenient building block for developers willing to
apply continuous integration methods to their HPC project without developing
their own launch scripts to speedup validation. As validation system reactivity is
a critical point, the important duration associated with large NRB can be a possible explanation of why some projects are not validated regularly. The purpose
of our work is to make HPC project validation suites more efficient in terms of
both computational costs and execution time. Indeed, a faster validation system
simplifying continuous testing opens the way for better programming practices
and transitively enhances code quality.

References

1. Cruisecontrol website. http://cruisecontrol.sourceforge.net/
2. TravisCI website. https://travis-ci.org/
3. Berg, A.: Jenkins Continuous Integration Cookbook. Packt Publishing Ltd,
   Birmingham (2012)
4. Calcote, J.: Autotools: A Practitioner's Guide to GNU Autoconf, Automake, and
   Libtool. No Starch Press, San Francisco (2010)
5. Dean, J., Ghemawat, S.: Mapreduce: simplified data processing on large clusters.
   Commun. ACM 51(1), 107–113 (2008)
6. Hoffman, B., Cole, D., Vines, J.: Software process for rapid development of HPC
   software using cmake. In: DoD High Performance Computing Modernization Program Users Group Conference (HPCMP-UGC), pp. 378–382. IEEE (2009)
7. Pérache, M., Jourdren, H., Namyst, R.: MPC: a unified parallel runtime for clusters of NUMA machines. In: Luque, E., Margalef, T., Benítez, D. (eds.) Euro-
   Par 2008. LNCS, vol. 5168, pp. 78–88. Springer, Heidelberg (2008). doi:10.1007/
   978-3-540-85451-7_9
8. Vavilapalli, V.K., Murthy, A.C., Douglas, C., Agarwal, S., Konar, M., Evans, R.,
   Graves, T., Lowe, J., Shah, H., Seth, S. et al.: Apache hadoop yarn: Yet another
   resource negotiator. In: Proceedings of the 4th annual Symposium on Cloud Computing, p. 5. ACM (2013)
9. Verma, A., Pedrosa, L., Korupolu, M., Oppenheimer, D., Tune, E., Wilkes, J.:
   Large-scale cluster management at Google with borg. In: Proceedings of the Tenth
   European Conference on Computer Systems, p. 18. ACM (2015)
10. Yoo, A.B., Jette, M.A., Grondona, M.: SLURM: simple linux utility for resource
    management. In: Feitelson, D., Rudolph, L., Schwiegelshohn, U. (eds.) JSSPP 2003.
    LNCS, vol. 2862, pp. 44–60. Springer, Heidelberg (2003). doi:10.1007/10968987_3

A Heterogeneous Runtime Environment for Scientific Desktop Computing

Nuno Oliveira[1,2](✉) and Pedro D. Medeiros[2](✉)

[1] ISEL - Instituto Superior de Engenharia de Lisboa, Lisboa, Portugal
no@deetc.isel.pt
[2] NOVA LINCS/Department of Informatics,
Universidade Nova de Lisboa, Lisbon, Portugal
pdm@fct.unl.pt

Abstract. Heterogeneous architectures encompassing traditional CPUs with two or more cores, GPUs and other accelerators like the Intel Xeon Phi, are available off the shelf at an affordable cost in a desktop computer. This paper describes work towards the definition, implementation and assessment of an environment that will empower scientists and engineers to develop and run their demanding applications in such personal computers. We describe HRTE (Heterogeneous Runtime Environment) that allows the construction of dedicated problem solving environments (PSE) taking advantage of those powerful and local processing elements, thus avoiding the use of remote machines through resource managers that introduce large latencies. HRTE is tailored to the communication and execution patterns of a PSE, efficiently mapping them to the heterogeneous architecture described. We also developed an API that eases the development of modules (*HModules*) that support multiple parallel implementations and are easily integrated in a traditional PSE.

HRTE functionality and performance and the API used to build *HModules* are assessed in the construction of a PSE in the area of Materials Science.

Keywords: Heterogeneous architecture · GPU · PSE (Problem Solving Environment) · Runtime environment · Accelerator · OpenCL

1 Introduction

Scientists have been conducted their research using increasing computational power to run their simulation models, to analyze large experimental data, and to compare observed and predicted data. The exploitation of the parallel hardware that supports the required levels of performance is too complex to one that is not a computer science expert. This complexity of hardware, middleware, software versions and standards must be hidden from the user. The objective is that an expert in a specific science could define his model or simulation without worrying about the runtime environment.

© Springer International Publishing AG 2017
I. Dutra et al. (Eds.): VECPAR 2016, LNCS 10150, pp. 256–269, 2017.
DOI: 10.1007/978-3-319-61982-8_23

Problem solving environments (PSE) are integrated environments for solving a target class of problems in an application domain. Typically, they encapsulate the state of the art algorithms and problem solving strategies through an easy interface in a way that an expert in the application domain could run his model without specialized knowledge of the underlying computer hardware or software. Several open source frameworks for building PSEs exist, namely OpenDX [14], Voreen [9] and SCIRun [11].

PSE environment offers the possibility of using building blocks from a library and interconnecting them in a network of modules that supports a dataflow model. The runtime of the PSE toolkit supports the dataflow between modules, the visualization of the intermediate and final results, as well as the modification of some parameters during the execution (steering computation). The network of modules can incorporate domain specific libraries such as numeric computation and visualization (see Fig. 1).

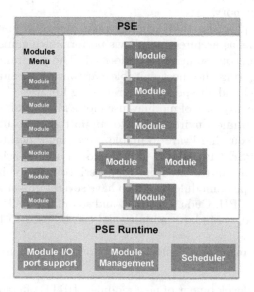

Fig. 1. A PSE environment with a network of modules.

A PSE provides a diverse set of modules with specific functions and the interface allows the user to build easily a network of modules. The execution of this network by the PSE runtime performs the processing steps needed to achieve the goal of the user. In each moment, the PSE scheduler determines the subset of modules that need to be executed according to data stream dependencies.

The runtime environments of PSEs need to support high requirements of computational power. This computational power is typically supported by cluster machines or even through the grid infrastructure. However, the use of remote parallel processing platforms implies the submission of requests through batch schedulers that introduce intolerable latencies for interactive use. A change in the technologies used for executing PSE modules is necessary in order to achieve a significant reduction of the processing times combined with small latencies that allows an interactive use by users. One promising way to achieve the above stated goal is through the exploitation of the heterogeneous multi-core architectures present in current desktop computers.

Thereby it is possible to develop new modules that take advantage of multiple CPU using frameworks as PThreads or OpenMP. For the same reason the operation of other processing units (PU), such as GPUs, can be carried out by the individual modules using frameworks like NVIDIA CUDA, OpenCL, etc. Therefore there is no obstacle in develop a module to take advantage of this type of hardware. These PUs are also known as accelerators and can share the main memory of the main processor (CPU) or having a private addressing space. In this work we used GPUs with its own separate address space. These types of PUs causes the module to explicitly copy the data into the memory of the PU, submit the code (kernel) to be executed, and finally copy of the data back to the main system memory.

The authors of [2] claim that in many cases the coordinated use of all the PUs of a heterogeneous architecture allows performance gains when a comparison with a homogeneous solution is performed. The programmer could implement modules targeting the most suitable hardware in mind, using a specific programming model and/or specific programming libraries.

To deal with the diversity of modules used for a given goal, we propose HRTE (Heterogeneous Runtime Environment) to support the execution of PSE tasks over the heterogeneous hardware available on a single desktop computer. We want to extend a PSE with HRTE in order to be able of schedule modules to the available hardware in the desktop computer. To follow this objective HRTE supports another type of modules that can have several implementations for each of the devices, e.g., CPU, Cuda, OpenCL, and so on. To the PSE user these new modules exist like any standard modules and can be used to build a processing network (see Fig. 2).

HRTE has two main parts that correspond to the two main contributes of this work:

- Simplifying the development of new modules. HRTE offers the notion of Heterogeneous Module (*HModule*) supporting several implementations for each type of PU allowing it to run on multiple hardware architectures (see Fig. 2). Support of transparent management of data copy between main memory and memory of the PUs is also included. The development of *HModules* is simplified through the availability of methods that implements map and stencil parallel control patterns [8] over HRTE.
- Optimizing the execution of the of module network: HRTE supports efficient access to large volumes of data flowing between modules in a complex memory hierarchy (including multicore CPUs, GPUs and other kinds of PUs). This data flow optimization between *HModules* is achieved through the minimization of the number of data transfers between CPU and PUs memories, taking advantage of the current location of the data.

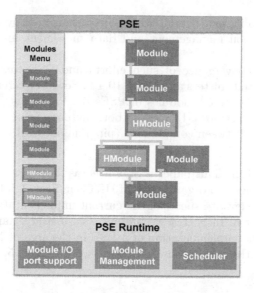

Fig. 2. A PSE environment with a network of modules and new HModules.

Related Work. Several research efforts that allow the exploitation of heterogeneous architectures for building efficient applications have been successful: OpenCL [6], HSA [5], StarPU [2], Harmony [4], and PTask [13].

OpenCL is a standard for cross-platform allowing the definition of kernels that can be offloaded to diverse processors found in personal computers as specific accelerators like GPUs and classical CPUs. OpenCL can be directly used to program a PSE's module.

Heterogeneous System Architecture (HSA) is a new hardware and software platform allowing software to use different types of processors, e.g. GPUs and CPUs, that work together through shared memory to efficient run demanding applications.

StarPU is runtime system providing a unified execution model together with a data management library. As claim by authors the main objective is to provide numerical kernel designers with a simplify way of defining task to run over the heterogeneous hardware. StarPU scheduler will assign task to available devices and the systems also allows the development and integration of new scheduling policies.

Harmony is a runtime supporting a programming and execution model providing the simplification of parallelism management, dynamic scheduling of kernels and monitoring the performance. Harmony supports compute kernels, analogous to imperative's function, that may have multiple implementations for multiples processors architecture. Execution of the kernels is determined by the set of input variables likes a dataflow approach.

PTask propose a set of operating system abstraction to support accelerators devices like GPUs as first class computing resources. These abstractions support a dataflow programing model built with objects managed by operating system allowing the execution management and data movement in the hands of the kernel.

Regarding the convergence of such efforts and PSE toolkits most of the projects have targeted clusters and grids [10,12]. Several references exist regarding the use of GPU-enabled modules in PSEs [7]. Parallel structured programming projects like FastFlow [1] address both heterogeneous architectures and support of dataflow between components (pipeline pattern).

Paper Organization. This paper is organized as follows: we begin by describing the characteristics and organization of HRTE in Sect. 2, followed by the presentation of some relevant aspects of the current implementation using SCIRun and StarPU in Sect. 3. In Sect. 4 we present a case study, namely the application of HRTE in the implementation of a PSE in the area of Materials Science. Finally we present the conclusions and future work in Sect. 5.

2   HRTE Organization

A PSE toolkit provides modules that can be interconnected with other modules in a dataflow approach. Each module reads data from its inputs, executes an algorithm and generates outputs to be sent to other modules. HRTE introduces a new type of module (*HModule*). These modules allow the execution of an algorithm in several platforms (hardware and software). The extensions should maintain compatibility with original features of PSE. Therefore all existing modules can still be used and can be interconnected with the new *HModules* (see Fig. 3).

In most PSEs large volumes of data are transferred between modules. The efficient support of these huge data transfers and the optimization of its sharing between modules must be tailored to an environment where a hierarchy of levels of memory exists; if we consider that some of the modules will be offloaded to an accelerator this implies that the data must also be transferred to and from the accelerators memory. The transfer costs must be considered by the runtime environment, otherwise the gains of using the accelerator can be hidden by the overheads intrinsic to data transfer between separate components of the memory hierarchy. Another issue is related with the limited memory in some accelerators, imposing that the accelerators memory may not accommodate all the data thus implying its partition. Therefore, HRTE must also extend the PSEs dataflow between *HModules* in order to send additional information about the locality and partitioning of the data transferred (see Fig. 3).

To be able to incorporate *HModules* in an existing PSE framework one must modify the PSE code to handle the execution of new *HModules* and the dataflow between both types of modules. The HRTE organization allows the minimization of changes of PSE code thus easing the integration of HRTE in different PSE toolkit. These modifications allow the definition of a new *HModule* by defining the methods summarize in Table 1.

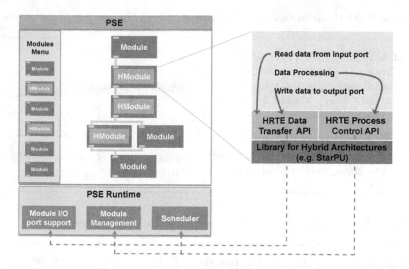

Fig. 3. A PSE environment with standard modules and the new *HModule* in same application.

Table 1. Methods need to be defined when declaring a new HModule

void getInputs()	Extract data from input ports and register it
void setOutputs()	Generate the data to the output ports of the module
void hexecute()	Definition of actions performed by the *HModule*

A dynamic library supports all the functionalities of the runtime and is used by the *HModule* code. It supports the concept of a heterogeneous function allowing, in the same module, the availability of different implementations. In Table 2, we present the method for adding an OpenCL implementation to a *HModule*.

Table 2. Method to register an OpenCL implementation in an HModule

void hrte_HFunction_add_opencl_code(
    hrte_HFunction *hf,
    char *kernelName,char *clFile)
	Register an OpenCL kernel implementation indicating the filename containing the OpenCL code.

The library also supports data management allowing data registration for on demand transfer between memory hierarchy levels and data partition (with and without ghost zones). The registration of the data in HRTE is done using the functions summarize in Table 3.

Table 3. Methods to register a 3D volume and set the number of partitions

```void hrte_matrix3d_register(```
```   hrte_data_handle *handle,```
```   void *ptr, uint32_t nx,```
```   uint32_t ny, uint32_t nz,```
```   size_t elemsize)```
	Register a 3D matrix.

```void hrte_matrix3d_set_partitions(``` ``` hrte_data_handle handle,int n)```	Set the number of partitions on data.

Table 4. HRTE map and the stencil parallel control patterns

```void hrte_task_map(hrte_HFunction *hf,```
```   hrte_data_handle in,```
```   hrte_data_handle out,```
```   hrte_HFunctionArgs *ha)```
	Map pattern will apply the heterogeneous function to every element of the input data.

```void hrte_task_stencil(hrte_HFunction *hf,```
```   hrte_data_handle in,```
```   hrte_data_handle out)```
	Stencil pattern will apply the heterogeneous function to every element and its neighbors.

To simplify the definition of a *HModule* map and the stencil parallel control patterns [8] are available and presented in Table 4.

3   Current HRTE Prototype

At present our prototype has been developed using SCIRun [11] as the PSE framework. As described in previous section we need to extend the SCIRun Module and the dataflow between modules to integrated HRTE and augmented SCIRun to support *HModules*.

The definition of a new module in SCIRun implies the definition of a new C++ class extending from the Module class and the definition of the virtual method *execute()* that is called when the module is executed.

We have the objective to minimize the intrusion in the source code of PSE and allows that all the existing modules will continue to be used and to be interconnect with the new *HModules*. So the integration of the heterogeneous module in SCIRun was done by defining a class *HModule* that extends the native class *Module* (see Fig. 4).

This class defines the virtual method *execute()* that will be called when the module starts its execution. In this case, the method will contain the HRTEs

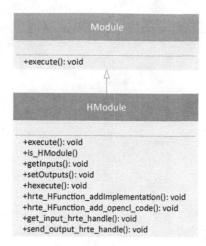

Fig. 4. UML class diagram representing the changes to the SCIRun source code for adding an HModule.

support code and finalizes calling the *hexecute()* virtual method. This approach allows the PSE framework to see *HModules* as native modules and their interface with the user, the management of module, data flows and the scheduling of modules is maintained as originally defined by the PSE toolkit.

Thus, the definition of a new HModule implies the development of a new C++ class that extends, not from the *Module* class, but instead from the *HModule* class. The programmer has to define the virtual method *hexccute()* instead of the method *execute()*.

The optional graphical user interface associated with the module is defined in TCL script language and finally the specification of the input and output ports are made in a XML file. The dataflow that interconnect modules was extended to include additional information when we have *HModules* interconnected.

The data binding with code with multiple implementations, support by HRTE runtime and described in Sect. 2, is carried out with the help of StarPU [2] environment. *HModules* are mapped to tasks and codelets; module input and output uses StarPU block management interface. The modifications made to the PSE dataflow part and the use of StarPU allows significant performance improvements when executing a sequence of *HModules*. This claim had been validated through the use of a network of *HModules* that runs an OpenCL kernel (Fig. 5) that only outputs the data received without any processing (void kernel).

The evaluation used a machine with an Intel Xeon CPU E5506 at 2.13GHz, 12 GB of RAM and two NVIDIA Tesla C1060. The operating system is Ubuntu 12.04 x86_64. The GPUs driver is the NVIDIA 340.29. The GPU SDK is CUDA 6.5.14 (OpenCL 1.1). PSE is SCIRun 4.7 and StarPU is 1.2.0rc2. In Table 5 we compare the execution times of this network with a similar one without HRTE (same OpenCL kernel). Optimization of dataflow between *HModules* allows a reduction in execution time up to 33% over the version that does not use HRTE.

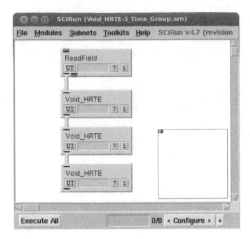

Fig. 5. Network used to evaluate the performance of dataflow between modules.

Table 5. Evaluation time of the performance of dataflow between modules with and without HRTE support.

Image size	Partitions	OpenCL	HRTE
100	1	29.3	23.6
	2	31.0	25.0
	4	27.5	26.1
200	1	55.2	31.6
	2	56.2	34.5
	4	67.7	31.4
300	1	200.9	53.3
	2	176.8	60.2
	4	205.6	62.4
400	1	545.3	188.3
	2	554.0	184.6
	4	556.3	188.5
500	1	1009.9	339.1
	2	1018.8	340.6
	4	1023.0	342.5
600	1	1772.7	575.6
	2	1768.8	574.1
	4	1785.8	578.1
700	1	2714.6	891.6
	2	2636.8	894.5
	4	2734.4	894.6

4   A Case Study in Materials Science

In the field of Materials Science, research on composite materials (comprising two distinct materials, where one constitutes a base matrix and the other acts as reinforcement) has a growing relevance in transportation and energy areas [3].

To forecast the characteristics of a new material, it is vital to characterize the reinforcements population regarding aspects such as position, size and orientation of the particles. X-ray computed tomography (X-ray CT) images are used for the characterization activities.

The task of processing and analyzing such data is a complex one: not only there is a huge volume of data to be processed but also there are noise and artifacts that must be removed; low contrast between the matrix and the reinforcement particles, due to small density difference makes this processing computing intensive.

Support of this processing and its easy handling by a non IT specialist requires an environment that allows the definition of a sequence of computational processing steps as well as its parameterization values in an interactive and real time way. In this setting, the construction of a PSE to the characterization of reinforcement population in 3D tomographic data is an opportunity for assessing the functionality and performance of HRTE. The images obtained by CT need processing to eliminate noise and allow the detection of boundaries between the base material and the reinforcement particles.

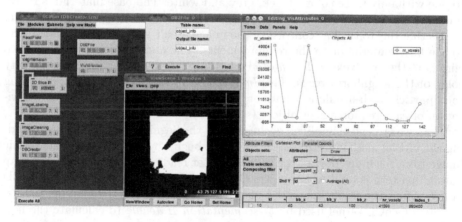

Fig. 6. PSE to characterize the reinforcements population determining its dimensions, position and orientation.

Figure 6 shows an example of the steps performed in order to obtain the characteristics of reinforcement population. The graphic in the right shows the particle's size distribution; to calculate the size of each reinforcement the PSE must segment the 3D image and this operation includes some steps that are computationally demanding. To assess HRTE, we ported an OpenCL existing

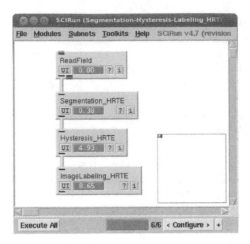

Fig. 7. Network to process CT raw images allowing the adequate objects identification.

implementation to HRTE. Both solutions have the same organization - in fact the OpenCL kernels are the same - that is described in the following.

We developed three *HModules* to process the tomographic 3D image. The modules perform in sequence *bi-segmentation*, *hysteresis* and *ImageLabeling* operations. *Bi-segmentation* transforms the CT 3D original grey scale image to one with only three colors: black, grey and white. The base material is represented as white, the reinforcements objects as black and the grey color represents voxels that due to the low contrast of the image aren't yet classified as belonging to the base material or to the reinforcements. The main goal of hysteresis is to eliminate the grey voxels. The *hysteresis* is implemented following the majority color of the neighbors voxels. The *ImageLabeling* segments the image labeling each particle with a unique identifier. This labeling step allows the characterization of each reinforcement object (see Fig. 7).

Next we present a simplified declaration of the Segmentation *HModule* including the virtual method *hexecute*. The method begins by reading the tomographic image from the input port of the module. After reading the input data, the map parallel pattern is used to apply the OpenCL kernel to all the voxels of the 3D image. After, the result image is sent to the output port of the module.

The OpenCL kernel used by *bi-segmentation HModule* to calculate the new value of all the image voxels is presented next:

```
__kernel void Segmentation (...){
BYTE value;
value = bk[INDEX(x, y, z, nx, ny)];
if (value <= min) value = BLACK;
else if(value > max) value = WHITE;
else value = GREY;
bkO[INDEX(x , y, z, nx, ny)] = value;
```

```
}

class Segmentation:public HModule {
    void hexecute() {
        hrte_data_handle img;
        get_input_hrte_handle(..., img, ...);
        ...
        hrte_task_map(...,img, img, ...);
        send_output_hrte_handle(..., img);
    }
};
```

Figure 7 presents the PSE network used in the experiment; Table 6 contains a table comparing the execution times of two networks built using the same approach described in Sect. 3: the 3rd column corresponds to a network where modules don't use HRTE support while the 4th column shows the execution time for the same kernels wrapped as *HModules*.

Table 6. Evaluations times of the network to process CT raw images executed with and without HRTE support.

Image size	Partitions	OpenCL	HRTE
100	1	59.6	57.2
	2	64.8	69.4
	4	66.3	70.1
200	1	218.6	175.7
	2	226.0	172.0
	4	228.1	180.9
300	1	709.9	512.9
	2	701.0	527.1
	4	725.1	532.1
400	1	1717.2	1193.4
	2	1724.4	1226.1
	4	1732.6	1240.8
500	1	3266.7	2323.6
	2	3274.5	2383.9
	4	3261.2	2403.9
600	1	5986.5	4297.8
	2	5984.0	4381.0
	4	6018.9	4424.7
700	1	10195.3	8255.1
	2	10209.3	8384.7
	4	10182.3	8466.8

5 Conclusions and Future Work

In this paper we presented HRTE which aims to ease the development of applications that use parallelism to tackle computational problems characterized by big needs in computational power and processing of big volumes of data. Our target is to help scientists and engineers that are not parallel processing specialists to develop modules for Problem Solving Environments toolkits that make an efficient exploitation of desktop computers equipped with accelerators. As far as we know, this effort to create a framework allowing the integration in Problem Solving Environments toolkits of modules that can have different implementations and communicate efficiently by optimizing the data transfers is original.

The case study in Materials Science showed that HRTE allowed that a network of already existing OpenCL-based modules executed taking advantage of the actual localization of data. On the other hand, details like data partitioning, device selection, and data transfer between different levels of memory hierarchy were hidden from the programmer.

The results obtained in our prototype using SCIRun and StarPU, assessed through a realistic 3D image processing are promising in terms of performance and also regarding the ease of development of new modules. The experiments described gave us valuable insights to further developments of our research efforts.

Our future work include the port to HRTE of other composite material processing modules and the assessment of HRTE in other areas of application. In our plans are also the support of *HModules* that support other parallel programming frameworks like MPI.

Acknowledgments. FCT MCTES and NOVA LINCS UID/CEC/04516/2013. The Polytechnic Institute of Lisbon (IPL) supports the 1st author as a doctoral student. FCT-funded Project PTDC/EIA-EIA/102579/2008 Tomo-GPU - Problem Solving Environment for Materials Structural Characterization via Tomography.

References

1. Aldinucci, M., Danelutto, M., Kilpatrick, P., Meneghin, M., Torquati, M.: Accelerating code on multi-cores with FastFlow. In: Jeannot, E., Namyst, R., Roman, J. (eds.) Euro-Par 2011. LNCS, vol. 6853, pp. 170–181. Springer, Heidelberg (2011). doi:10.1007/978-3-642-23397-5_17
2. Augonnet, C., Thibault, S., Namyst, R., Wacrenier, P.A.: StarPU: a unified platform for task scheduling on heterogeneous multicore architectures. Concurr. Comput. Pract. Exper. **23**(2), 187–198 (2011). http://dx.doi.org/10.1002/cpe.1631
3. Cadavez, T., Ferreira, S.C., Medeiros, P., Quaresma, P.J., Rocha, L.A., Velhinho, A., Vignoles, G.: A graphical tool for the tomographic characterization of microstructural features on metal matrix composites. Int. J. Tomogr. Stat. **14**(S10), 3–15 (2010)
4. Diamos, G., Yalamanchili, S.: Harmony: an execution model and runtime for heterogeneous many core systems. In: HPDC08. ACM, Boston, Massachusetts, USA, June 2008

5. Hwu, W.M.W. (ed.): Heterogeneous System Architecture: A New Compute Platform Infrastructure. Morgan Kaufmann Publishers Inc., Waltham (2016)

6. Khronos: OpenCL (2016). http://www.khronos.org/opencl

7. Leeser, M., Yablonski, D., Brooks, D., King, L.S.: The challenges of writing portable, correct and high performance libraries for GPUs. SIGARCH Comput. Archit. News **39**(4), 2–7 (2011). http://doi.acm.org/10.1145/2082156.2082158

8. McCool, M., Reinders, J., Robison, A.: Structured Parallel Programming: Patterns for Efficient Computation, 1st edn. Morgan Kaufmann Publishers Inc., San Francisco (2012)

9. Meyer-Spradow, J., Ropinski, T., Mensmann, J., Hinrichs, K.: Voreen: a rapid-prototyping environment for ray-casting-based volume visualizations. IEEE Comput. Graph. Appl. **29**(6), 6–13 (2009). http://dx.doi.org/10.1109/MCG.2009.130

10. Miller, M., M.C.D.J.J.C.: Grid-enabling problem solving environments: a case study of SCIRun and NetSolve. In: Proceedings of HPC 2001, pp. 98–103, 22–26 April 2001

11. Parker, S.G., Johnson, C.R.: SCIRun: a scientific programming environment for computational steering. In: Proceedings of the 1995 ACM/IEEE Conference on Supercomputing. Supercomputing 1995. ACM New York (1995). http://doi.acm.org/10.1145/224170.224354

12. Peterson, J., Hallock, M., Cole, J., Luthey-Schulten, Z.: A problem solving environment for stochastic biological simulations. In: Proceedings of the 3rd Workshop on Python for High-Performance and Scientific Computing (2013)

13. Rossbach, C.J., Currey, J., Silberstein, M., Ray, B., Witchel, E.: PTask: operating system abstractions to manage GPUs as compute devices. In: Proceedings of the Twenty-Third ACM Symposium on Operating Systems Principles. SOSP 2011, pp. 233–248. ACM, New York (2011). http://doi.acm.org/10.1145/2043556.2043579

14. Thompson, D., Braun, J., Ford, R.: OpenDX: Paths to Visualization, 1st edn. Visualization and Imagery Solutions Inc, Missoula (2001)

Author Index

Printed in the United States
By Bookmasters